THE TEST AND ITCHEN ASSOCIATION

CHALK STREAMS

A Guide to their Natural History and River Keeping

To mark the Centenary of the Test and Itchen Association

Editor

Jim Glasspool

First published in 2007
by the Test and Itchen Association Ltd
West Haye, Itchen Abbas, Winchester SO21 1AX

ISBN 978-0-9555761-0-2

Typeset and printed by
Sarsen Press, Winchester

ACKNOWLEDGMENTS

In producing this book to mark the Centenary of the Test and Itchen Association, I should like to acknowledge the debt which is owed to the authors and those who have freely provided illustrations. Without their support and encouragement, this book could certainly never have been produced. I trust that they are all individually recognised sufficiently in the book.

I should also like to pay due recognition to Roy Darlington whose original manuscript was the inspiration for the book and provided much of the material for Chapter I.

I would also like to thank Lawrence Talks for all the work he has done on the provision of material, editing and generally encouraging the editorial group.

Finally I would like to thank Judith Blake who has taken a mass of text and illustrations in a variety of different forms and created a design and layout which is truly worthy of the subject matter and the occasion for which it has been produced.

Jim Glasspool

CONTENTS

Foreword 1
Jim Glasspool, *Editor*

Chapter One – Origins and Evolution 3

Geological history
Afforestation 6
Early human inhabitants 6
The first engineers 7
The Middle Ages 7
The meadsmen and drowners 8
The decline of the water meadows 9
The development of fishing 10
The growth of regulation 12

Chapter Two – Aquatic Plants 17
Michael Baron, *formerly Head of Science, Winchester College*

Conditions for growth 17
Adaptations of plants 17
The value of the plants 18
Problems 18
Identification of aquatic plants 19
Important submerged water plants 19
Table of submerged plants 25
The algae 27
The variety of algae 27
Seasonal changes in algal growth 27
Control of algal growth 28
Management of aquatic plants 29
Growth and control 30
Value of different aquatic plants 30
Aquatic and marginal vegetation 31
Emergent vegetation 31
The river margin 31

Chapter Three – The Ecology of the River Bank 37
Michael Baron, *formerly Head of Science, Winchester College*

The riverine and bankside habitat 37
Importance 37
Range of habitats 37
Vegetation of the riverside 37
Table of plants found on the riverbank 38

Alien species 46
Watercress 47
Cut paths and grassland 47
Sedge and reed beds 48
Fenland meadows 49
Carr woodland 49
Animal life of the riverside 50
Insects 50
Birds 51
Mammals 52
References 53

Chapter Four – Chalkstream Invertebrates 55

Warren Gilchrist, *Fellow of the Royal Entomological Society*

The river invertebrates 55
The most significant foods of game fish 56
Sensitivity to pollution 56
Life habits and requirements of the most significant invertebrates 56
The upwinged flies 56
The nymph, or larva 58
The dun, or sub-imago 59
The spinner, imago or perfect insect 59
The eggs 59
Identification of the Ephemeroptera 60
The identification process 60
Identification of the nymphs 61
Identification of the duns, or sub-imagines 61
The Ephemeroptera duns which are self-evident 62
The Ephemeropteran duns which are not self-evident 67
Key to Ephemeropteran duns with two tails 68
Identification of the spinners 69
Spinners which are self-evident 69
Spinners which are not self-evident 70
Key to Ephemeropteran spinners with two tails 70
The sedge or caddis flies (Trichoptera) 71
The true-flies (Diptera) 75
Reed smuts and black-flies 75
Gnats 75
Non-biting midges 75
Crane-flies 76
The Hawthorn Fly 76
Crustaceans 77
Freshwater shrimps (Amphipoda) 77
The hog-lice or water slaters (Isopoda) 77
Crayfish (Astacidae) 77
Snails, limpets and mussels (Gastropoda and Bivalvia) 79
Other invertebrates 81
Times of emergence 85
Conservation of invertebrates in a river 85

Aquatic plants 85
The nature of the river bed 85
Fly boards and flints 86
Sites of Scientific Interest (SSSIs) 87
Habitat 87
Electro-fishing and fly life 88
Channel and bank works 88
Pollution 88
The abundance of fly life 89
Records and recording 90
Appendix A Checklist of the British Ephemeroptera 91
Appendix B References 92
Appendix C The Amended Biological Monitoring Working Party (BMWP) score system for sensitivity to pollution 93
Appendix D Simplified Invertebrate monitoring for anglers 94
Appendix E Times of emergence of the Ephemeroptera 97

Chapter Five – Chalk Stream Fish and their Management 99
Shaun Leonard, *Head of Fishery Studies, Sparsholt College*

The fish of the chalk streams 99
Brown Trout, *Salmo trutta* 99
Atlantic Salmon, *Salmo salar* 102
Rainbow Trout, *Oncorhynchus mykiss* 104
Grayling, *Thymallus thymallus* 105
Pike, *Esox lucius* 106
European Eel, *Anguilla anguilla* 107
Other fish species 110
Stock Management 112
Management of wild stocks 113
Improving egg eurvival 113
Improving fry survival 115
From parr to adult 117
Salmon fisheries 119
Regulation of fishing 119
Stocking fish 119
Species stocked 120
Brown Trout 120
Rainbow Trout 121
Stocking practices 121
Size of fish to stock 122
Important details in stocking 123
Holding facilities 123
Fish disease 125
The control of predators and competitors 126
Mammals 127
Otter 127
Mink 127
Other mammals 128

Birds 128
Fish 129
Pike 129
Eel 129
Grayling 130
Methods of control of unwanted fish species 130
Angling 130
Netting 130
Fish traps 131
Other methods 131
Electric fishing 131
Effects of electricity on fish 132
Effects of electricity on invertebrates 132
Field use 133

Chapter Six – The River Keeper's Year 135
Guy Robinson, *Head River Keeper, Leckford Estate*

Winter work 135
Leaving the fringe 135
Managing grayling and pike 135
Autumn weed cut 136
Cleaning spawning gravels 136
Ditch clearance and tree management 136
Vermin 137
River bank maintenance 137
Planning – Need to obtain consent 137
Timing 137
Traditional river bank repairs 138
Mudding 140
Punts 140
Bridges 141
Fences 142
The fishing hut 142
Hatches 143
Stiles and gates 144
Fishing seats 144
Summer work 145
Fish stocking 145
Weed cutting 146
River keepers and aquatic plants 148
Technique 149
Line and bar cutting 149
Pole scythe 151
Chain scythe (links) 151
Weed cutting boats 152
Weed jams 153
Weed racks 153
Herbicides 153

Planting weeds 153
Trees 154
Mowing 154
Fishing instruction 155
General Section 155
Poaching 155
The water bailiff 155
The river keeper 155
Beat markers 156
Fishery rules 157
Fishery records 157
Equipment 158
Essential hand tools 158
Winches 158
Portable electric winches 158
Grass cutters 158
Brush cutters 159
Brush cutter safety 159
Chain saws 159
The workshop 160
References 161

Retrospect 162

LIST OF ILLUSTRATIONS

Illustration Number **Page**

Chapter One

1 Chalk rivers of England 2
2 Extent of chalk strata in Southern England 4
3 The Solent River system 5
4 Vitruvius mill 7
5 Diagram of a typical water meadow system 9
6 Window in Silkstead Chapel, Winchester Cathedral 10
7 Silkstead Chapel with tomb of Izaak Walton 11
8 'A Member of the Houghton Fishing Club' by J.M.W. Turner 12
9 River Test, before and after dredging 13
10a, b Eutrophication and low flow causing algal growth and loss of weed on the Itchen 15

Chapter Two

11 The rich and varied life of a chalk stream 16
12 Well managed chalk stream 18
13 Water-crowfoot *Ranunculus penicillatus* 20
14 River Water-dropwort *Oenanthe fluviatilis* 21
15 Horned Pondweed *Zannichellia palustris* 21
16 Water-starwort *Callitriche* sp. 22
17 Lesser Water-parsnip *Berula erecta* 22
18 Fool's Watercress *Apium nodiflorum* 23
19 Mare's -tail *Hippuris vulgaris* 23
20 Canadian Pondweed *Elodea canadensis* 24
21 Willow Moss *Fontinalis antipyretica* 24
22 Azolla Fern *Azolla filiculoides* 29
23 Long skeins of algae *Cladophora* 29
24 Watercress *Rorippa nasturtium-aquaticum* 32
25 Water-speedwell *Veronica anagallis-aquatica* 32
26 Water Forget-me-not *Myosotis scorpioides* 33
27 Water Mint *Mentha aquatica* 33
28 Hairy Willowherb *Epilobium hirsutum* 33
29 Purple-loosestrife *Lythrum salicaria* 34
30 Orange Balsam *Impatiens capensis* 34
31 Monkeyflower *Mimulus guttatus* 34
32 Bur-reed *Sparganium erectum* 34
33 Greater Tussock-sedge *Carex paniculata* 35
33a Common Reed *Phragmites australis* 35

Chapter Three

34 Barn owl quartering the water meadow 36
35 Reed Sweet-grass *Glyceria maxima* 40
36 Marsh Marigold *Caltha palustris* 40
37 Comfrey *Symphytum officinale* with Hogweed in foreground 40
38a, b White and Purple Comfrey *Symphytum officinale* 41
39 Flag Iris *Iris pseudacorus* 41
40 Meadow-sweet *Filipendula ulmaria* 42
41 Marsh Woundwort *Stachys palustris* 42
42 Water Dock *Rumex hydrolapathum* 42

Illustration Number		Page
43	Fleabane *Pulicaria dysenterica*	43
44	Hemp Agrimony *Eupatorium cannabinum*	43
45	Water Avens *Geum rivale* amongst Reed Sweet-grass	43
46a, b	Hemlock Wate-dropwort *Oenanthe crocata*	44
47	Marsh Orchid *Dactylorhiza praetermissa*	45
48	Figwort *Scrophularia nodosa*	45
49	Common Meadow Rue *Thalictrum flavum*	45
50	Yellow Loosestrife *Lysimachia vulgaris*	46
51	Ragged-Robin *Lychnis flos-cuculi*	46
52	Himalayan Balsam *Impatiens glandulifera*	46
Chapter Four		
53	Mayfly, *Ephemera danica*, male spinner	54
54	Chalk stream larvae	57
55	Mayfly, *Ephemera danica*	63
56	Blue-winged Olive, *Serratella ignita*	63
57	Yellow May nymph, *Heptagenia sulphurea*	63
58	Mayfly, *Ephemera danica*, female dun	63
59	Blue-winged Olive, *Serratella ignita*, female spinner	64
60	Iron Blue dun, Alainites muticus, male dun	64
61	Turkey Brown, *Paraleptophlebia submarginata* female dun	64
62	Large Dark Olive, *Baetis rhodani*, female dun	65
63	Pale Watery, *Baetis fuscatus*, male dun	65
64	Small Dark Olive, *Baetis scambus*, male dun	65
65	Large Spurwing, *Procloeon pennulatum*, female dun	66
66	Small Spurwing, *Centroptilum luteolum*, male dun	66
67	Pond Olive, *Cloeon dipterum*, female dun	66
68	*Micropterna sequax*, Limnephilidae	73
69	Grannom, *Brachycentrus subnubilus*, larvae	73
70	*Plectrocnemia conspersa*, Polycentropodidae	73
71	Grannom, *Brachycentrus subnubilus*, adult	73
72	Mosquito gnat, *Culiseta annulata*	73
73	Chironomid fly, *Cricotopus sylvestris*	73
74	Small Crane-fly	76
75	Crane-fly, *Tipula paludosa*	76
76	Hawthorn Fly, *Bibio marci*	76
77	Freshwater shrimp, *Gammarus*	77
78	Freshwater Louse, *Asellus aquaticus*	77
79	White-clawed Crayfish, *Austropotamobius pallipes*	78
80	Signal Crayfish, *Pacifastacus leniusculus*	78
81	Pond Snail, *Lymnaea peregra*	79
82	Great Ramshorn Snail, *Planorbarius corneus*	79
83	Shining Ramshorn Snail, *Segmentina nitida*	80
84	Pea Mussel, *Pisidium* sp.	80
85	Southern Damselfly, *Coenagrion mercuriale*	80
86	Banded Demoiselle, *Calopteryx splendens*, nymph	80
87	Banded Demoiselle, *Calopteryx splendens*, adult	80
88	Greater Water Boatman, *Notonecta glauca*	82
89	Greater Water Boatman, *Notonecta glauca*	82
90	Alder fly, *Sialidae*	82
91	Stonefly, *Isoperla grammatica*, nymph	82
92	Stonefly, *Perlodidae*, nymph	82
93	'February Red' Stonefly, *Taeniopterix nebulosa*, adult	82
94	Fly boards provide a place for invertebrates to lay their eggs	86

Illustration Number **Page**

Chapter Five

95 Kingfisher 98
96 Trout parr 100
97 Good quality stocked brown trout 101
98 A wild brown trout 101
99 Salmon: four stages of development 102
100 Salmon parr 102
101 Juvenile salmon and trout, showing their obvious differences 103
102 A wild British rainbow trout parr 104
103 Grayling 106
104 The view down the throat of a large pike 107
105 European eels 108
106 A large 'yellow' eel 109
107 Historic eel traps on the Test at Leckford 109
108 Bullhead 110
109 Bullhead eggs attached to the underside of a submerged stone 111
110 The ammocoete larva of the brook lamprey 111
111 An adult brook lamprey 112
112 A cleaned patch of gravel 114
113 In-stream incubators or 'egg boxes' can provide a safe hatching environment and salmon eggs 115
114 Newly hatched trout 116
115 A large flint is placed on a newly restored spawning riffle 116
116 Raceways or 'stews' for growing on restocking trout 124
117 This wound was probably caused by a heron 126
118 Bite marks and scratches on a carp 127
119 Game Conservancy Trust mink raft 128

Chapter Six

120 Martin Murrell, Chairman Hampshire River Keepers, with the tool kit for today's river keeper 134
121 Gravel cleaning 136
122 Posts and willow spiling are used to repair an eroded bank 138
123 Hazel faggots are used to create the outline of an island 139
124 Chalk back-fill consolidates the island 139
125 Top soil is spread over the chalk 139
126 Footbridge over the River Test 141
127 Guy Robinson, Leckford Head River Keeper 143
128 Adjustable hatch 144
129 Leckford fishing seat complete with notch to rest your fishing rod 145
130 Fish transporter 146
131 Fish boxes are used to stock trout along the river 146
132 Paul Moncaster clearing down after weedcutting on the Test at Tufton 148
133 Weed cutting with Turk scythes 150
134 A pole scythe being used to cut weed in deep water 150
135 Weed cutting boats are used in deep water 152
136 A chainsaw being used to cut willow faggots 160

Retrospect 162

137 Upper Test at Polehampton 163
138 Itchen water meadows at St Cross 163
139 Upper Test at Tufton 164
140 Middle Test at Bossington 164

FOREWORD

The chalk streams of Southern England have long been recognised as some of the finest examples of their kind in the world, with the Test and Itchen being described as 'The Jewels in the Crown'. The rural nature of the river valleys, the diversity of species and the high quality of the game fishing have made them internationally famous. Indeed much of the early history of game fishing took place on these rivers and the literature testifies to their importance in the development of modern fly fishing practice.

Perhaps less well known is the extent to which the management of these rivers by man over the centuries has shaped and protected their environment. The rivers have come a long way from their original primeval state but at the same time have maintained a natural environment, rich in species and having a distinct and beautiful landscape.

Those who currently care for the chalk streams are the inheritors of two thousand years of river control and management. The techniques and practices which characterise current management represent an amalgam of the trial and error of past centuries' experience, overlaid with more recent scientific knowledge.

An understanding of both the geological and the human history of the rivers is essential to an appreciation of the current state of the rivers and the regime which is required to conserve their ecology and landscape.

From the major engineering works of the Romans, who probably built the first water mills, through the now ghostly figures of the water meadow drowners of the eighteenth and nineteenth centuries, to the water keepers of the last hundred years, human beings have shaped the rivers and made them what they are today.

This historical tradition is now supplemented by scientifically based knowledge which is being added to all the time by organisations such as the Environment Agency, the Institute of Fisheries Management, the Institute of Freshwater Ecology and the many universities and colleges which are studying all aspects of river life. In Hampshire, Southampton University and Sparsholt College are carrying out valuable research projects and the latter is also training future generations of fishery managers who will have a more formal training and a better understanding of the scientific basis than their predecessors.

Because of the way in which the Test and Itchen have been shaped by man, they depend today on continued maintenance and conservation by the river keepers. It is largely the fishing interest which pays for this care and attention and which conserves the wildlife habitat. Without the income from fishing it would not be possible to employ the keepers who maintain the banks and there would not be the same incentive for owners to preserve the rural character of the valley. Other influences, particularly development pressures, would take over. However, it is most important that this essential care and maintenance is carried out in a way which conserves and enhances the many facets of the river environment. The continued conservation of this rich environment is the key theme of this book.

It is hoped that the book will be of interest not only to riparian owners and the river keepers responsible for the management of the rivers, but also to all those who fish the rivers, observe its wild life or are generally interested in the wild life habitat and its conservation.

Jim Glasspool

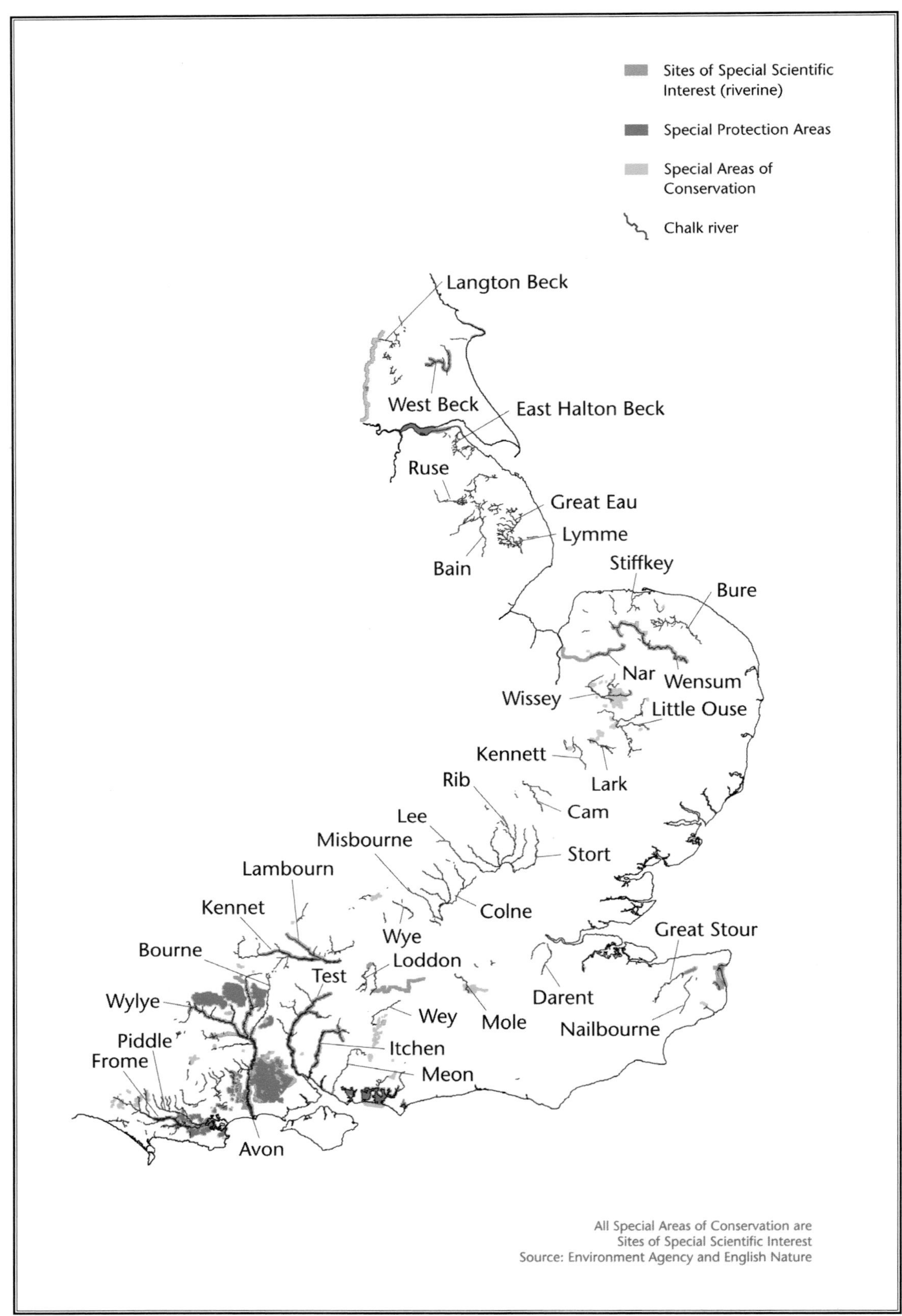

1 Chalk Rivers of England

Chapter One

ORIGINS AND EVOLUTION

When related to the many thousands of miles of rivers and streams which drain these islands, the chalk streams represent a tiny proportion of the total extent. However their significance in terms of the place they occupy in angling literature cannot be over estimated. It is no mere coincidence that some of the great innovators of game fishing practice, from Izaak Walton onwards, found their inspiration beside these rivers. Notwithstanding their proximity to London, the streams possessed the ideal characteristics of clarity, stability and biotic potential which rendered them the perfect medium for the studies and inventions of these great anglers. To understand the unique features which made these streams so ideal for their work it is helpful and interesting to consider those factors, both natural and contrived, which have influenced the origin and evolution of these watercourses.

Geological History

In the warm clear seas which covered what is now Southern England, during the latter half of the Cretaceous period, between 70 and 88 million years ago, there was a proliferation of countless millions of bi-valves and minute marine plants such as coccoliths. The sea floor received immense quantities of the shells of these creatures, which together with the skeletal parts of the decomposing coccoliths, became compressed to form the vast deposits of chalk, a material consisting almost entirely of calcium carbonate. Deposition completed, the Cretaceous period ended about 65 million years ago, bringing to an end the Mesozoic era. The transition into that section of geological time known as the Tertiary period was marked by a major recession of the sea, leaving extensive outcrops of the younger chalk, some of which were in excess of 1600 feet deep. Although comparatively minor fluctuations of sea level followed, the water never rose to a level which would submerge the upper chalk.

The earth movements which gave birth to the Alpine mountain ranges, took place about 15 million years ago. Tremendous shock waves emanating from the hub of this activity, buckled and folded the chalk strata in a series of waves and troughs along an east-west axis. These movements completed the structural pattern of the region. Today the waves of this structure are represented by the highest points such as Dean and Portsdown Hill which are both located on the Portsdown anticline. In East Hampshire, Sussex and Kent the buckling and subsequent erosion led to the formation of the Weald. So it was that time and the cataclysmic forces of nature combined to create the immense catchment area now formed by the rolling downland. Precipitation, particularly in the winter, soaks through the thin layer of topsoil to be absorbed and held within the cool depths of the porous chalk, which months later liberates this water through springs located lower down, resulting in a more constant flow regime than in spate rivers. This subterranean region of saturated chalk is known as the 'aquifer'.

2 Extent of chalk strata in Southern England

In Britain a series of glacial epochs began a little over half a million years ago, heralding the arrival of the Pleistocene period. It was an era of considerable climatic fluctuation. The ice ages, the most recent of which began about 75,000 years ago, lasting until almost 9,000 B.C., were periods of intense glaciation and subsequent thaw, producing long term modification of drainage patterns due to varying rates of erosion. They also caused associated rises and falls in sea level which were responsible for the final disruption of the Solent river system, about 8,000 years ago, when the sea level rose by 45 metres. It is thought that a major river drained East Dorset, Wiltshire and Southern Hampshire, to join the sea at some distance off what is now Selsey Bill. The Test and Itchen were merely tributaries of the Solent river, as were the upper Avon,

Wylye and Nadder in Wiltshire and the Frome and Stour systems in Dorset. The Isle of Wight would not have existed as an island because of the much lower sea level. Since it was first formed more than 230,000 years ago the Solent river system would have been disrupted numerous times by changes in the sea level.

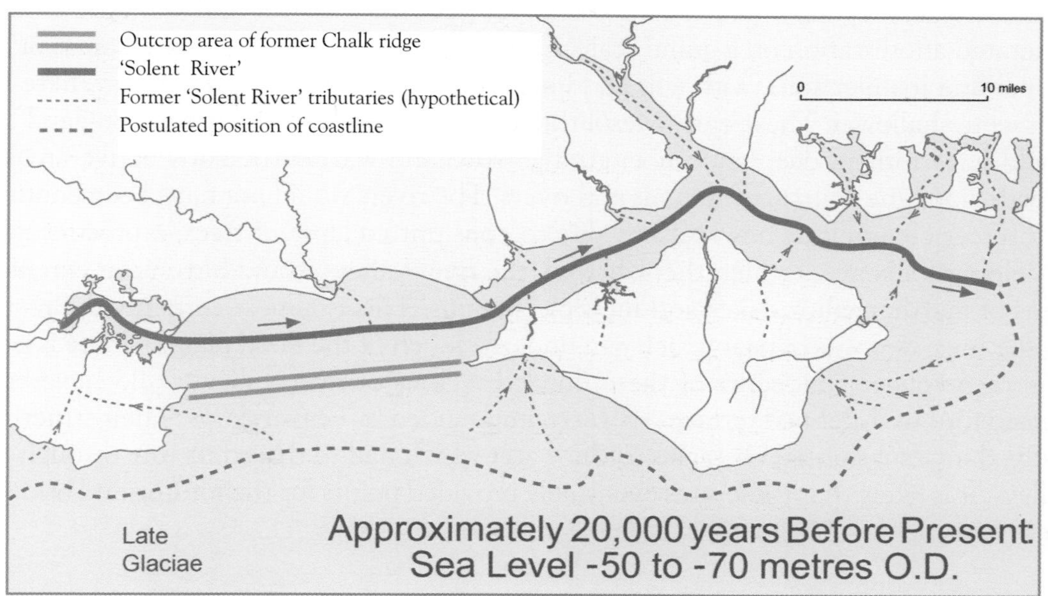

3 The Solent River System *Scopac/Portsmouth University*

A further characteristic of the Hampshire basin is the extensive area of dry valleys. These were almost certainly formed during the Pleistocene period with its protracted series of freeze and subsequent thaw. The vast ice sheets which dominated the landscape to the north did not extend further south than the Midlands. The slightly warmer conditions which prevailed in the south meant that a huge volume of melt water had to drain seaward through the southern counties and particularly across the Hampshire basin. Although surface icing was not as pronounced as further north, water was frozen at some depth within the chalk, causing it to be impermeable and thereby able to support vast streams of water running from the thawing ice. A warming climate towards the end of the Pleistocene period produced enormous spring-flows. These combined with the frozen ground, caused an accelerated rate of headward erosion, especially in the upper reaches of all the rivers and streams. When the thaw finally ended the dry valleys and combes remained, evidence of what must have been a remarkably bleak landscape.

Another interesting, if complex, feature of this period is the continuous and progressive modification of drainage patterns. Because of fundamental differences in the composition of the chalk strata, there were differing rates of erosion. Even neighbouring streams could have varying speeds of headward erosion. Sometimes a tributary of such a pair of rivers would cut back progressively towards its neighbour, subsequently intercepting its course and often diverting the entire flow away from the original channel, a geological process known as 'River Capture'. An example of this was the capture of the upper reaches of the Candover Stream, a watercourse which would have been quite substantial, running into the sea along the course of what is now the much reduced river Meon. For structural reasons the rate of downward erosion of this stream was much slower than that at which a tributary of Itchen was cutting its way eastward. The inevitable interception took place, resulting in a drainage pattern much as it is at present.

Afforestation

The last glacial epoch came to its close about 9000 B.C. A wet and warming climate generated afforestation on a grand scale. Oak and hazel thickets dominated areas of the deeper and loamier soils, with elm and lime spreading throughout the areas where the soils were shallower. These early Mesolithic forests spread across Southern England and would have formed deep and impenetrable woodland with particularly active areas of growth along the courses of streams and rivers. The rivers would not have been confined to discreet channels as now. Rather, due to constant collapse of trees, a process aided by colonies of beavers which then thrived, the river valleys would have been extensive areas of marsh, shallow lakes and multiple streams. This is almost certainly the reason for the huge deposits of peat which can underlie much of the flood plains. There is good reason to believe that some of these flooded regions of river valley would have been formed into well-defined terraces of water, impounded by beaver dams. Their structures evolved into substantial obstacles to the water which had to thread its way through the valleys. It is likely that these sites eventually provided points for the fording and bridging the rivers and possibly later, ready-made mill sites.

Early human inhabitants

In Neolithic times the country would have presented a picture of endless forest canopy with rivers draining through numerous undefined channels, diffusing into extensive shallow lakes and insect infested swamps. These would have been dangerous places for stone age man and his dwellings would have been confined to the higher clearings. He probably would have made only brief excursions to the river to draw water and possibly to fish, although there is little evidence of the methods that were used. There are, however, archaeological indications that the shores of Southampton Water were frequented for fishing.

The Bronze and Iron Ages marked periods when the hill fort communities became better organised and more sophisticated. Excavations at Danebury reveal a picture of quite advanced farming practices, with chalk pits being used to store huge quantities of grain. Sheep grazed the open areas and pigs grubbed the woodland fringes, indicating substantial clearance of the forest areas had already taken place. These Iron Age people were essentially Celtic in origin and many of the local place names reflect this influence. Winchester was 'Caer Gwent' (White City) before it became Venta Belgarum, and the Celtic names for a spring, 'an' or 'aen' is reflected in the ancient names 'Terstan' and 'Icenan' for the rivers Test and Itchen. Similarly, the Candover and Dever rivers take their name from the pre-Roman British Celtic 'Caniodubri' meaning 'beautiful waters', and 'Micnodubri' meaning 'bog waters', perhaps from the fact that the river drained into marshes on the Test.

During the Iron Age more and more land was cleared of trees, settled, cultivated and grazed. The mainstay of agriculture was corn-growing and it is fairly certain that the grain was ground laboriously by hand using crude saddle stones. The first rotary quernstone was a significant advance – a development which probably occurred before the advent of the Romans in AD 43.

The first engineers

With Roman military expertise came the Roman engineers, men who knew how to make improved weapons, agricultural implements and of course roads. They engineered a sophisticated network of roads, not only to administer their garrisons, but also to service the developing farming communities. With increasing yields of corn, it was the Roman engineers who brought the rivers under control and installed the first watermills. A well documented design by Vitruvius had a vertical wheel and a mechanism astonishingly similar to those mills which survive today. Roman Hampshire would have provided ideal conditions for the establishment of mill sites and Winchester or 'Venta Belgarum' was a thriving centre for trade. It seems likely that before the Romans departed they would have engineered much of the drainage, waterworks and mill sites, the pattern of which survives to this day.

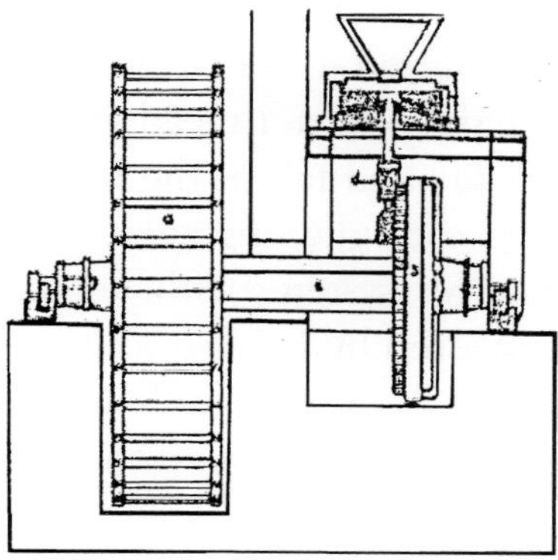

4 Vitruvius mill

Excavation of Iron Age hill forts has revealed considerable quantities of fish bones and it seems likely the rivers provided a valuable food source. The use of fish traps, normally made from osiers(willows), is well proven, and the early mill sites would have lent themselves conveniently to the operation of eel traps. The containment of watercourses within fairly well-defined channels as a result of Roman engineering would have made it much easier to string ranks of fish traps across the rivers.

Little is known of Saxon attitudes to river works and fisheries, although it is reasonable to assume that the provision of food from local streams was important and they would have needed to maintain the watercourses associated with the mills.

The Middle Ages

Twenty years after the Norman Conquest, the Domesday Survey recorded over 300 mills in Hampshire and the Isle of Wight. It also records a fishery valued at threepence at Houghton, while one at Otterbourne was worth two shillings. Salmon were prolific and salmon fisheries were valuable assets. The notable salmon fishery at Woodmill tidal pool was recorded in 1275 as having salmon fishings with an annual value of ten marks.

The founding of the monasteries probably brought the art of fish-culture to these islands, because fish, mainly carp, figured greatly in the diet of monks. Fishponds were created by damming streams and many of these early techniques of pond construction and water control are still in use today. Stewponds were not confined to the monasteries, since many secular manors boasted them too. A number of these ponds remain today such as those at Foley and Bohunt, while those at Southwick, Beaulieu and Mottisfont belonged to religious orders. New water courses and feeder streams would have been necessary to serve the stewponds and these were yet another feature permanently modifying the essential structure and appearance of the valleys.

The meadsmen and drowners

Apart from the impact of water mills and the construction of fishponds on the valleys, one further development was to come which would substantially modify the pattern of drainage in the valleys.

Around the beginning of the 17th century, the river Wey witnessed the inauguration of a primitive system for improving the drainage and irrigation of alluvial meadows. By 1618, the start of a basic system at Bishopstoke on the Itchen was in evidence. By the end of the 18th century many thousands of acres of wet valley pastures, considered agriculturally poor, had been converted to water meadows.

Essentially there were two types of water meadow. Those that could be flooded and drained using the natural contours and fall of the land, and those which, because the valley floor was flat, required a carefully engineered system of carriers and drainage channels. In Hampshire, the latter system, with few exceptions, is the predominant method used. Some were engineered by English and some by Dutch experts. Most of the water meadow systems on the Avon were of Dutch construction, often considered more efficient than their English counterparts, as the Dutch system made better use of any natural level advantages.

The water meadows consisted of a series of ridges 4-5 metres apart and about 1 metre high. Water was diverted from the river along a carrier and then along smaller 'drowners' or 'floaters' on the top of the ridges. Using oak hatches on the carriers and turves to stop the flow on small carriers, the water level was controlled so that it overflowed the 'drowners' and floated the strips of adjacent meadow known as 'panes'. Small drains were dug between the drowners, which took the water back to a main drain which carried it back to a point further down the river or side stream which supplied it. It was important to maintain a constant flow of water over the panes at a depth of no more than 1-2 inches. This required considerable skill on the part of the meadsmen or drowners who operated and maintained the system. A gradient of 1 in 200 was considered the optimum and this was achieved using only simple hand tools. It was important that the meadows were not allowed to become waterlogged, as this promoted the growth of rushes and reeds.

The operation of the water meadows was linked to the system of sheep and corn on the chalk downland. Sheep were kept on the downlands to manure the thin chalk soil to give a good cereal crop. But the size of the flock was limited by the amount of feed available in the winter and early spring.

By flooding the meadows two or three times in January and February, the warmer river water increased the soil temperature, producing an early flush of growth or 'early bite' which could be used for grazing in March, before the downland grazing was available. Grazing ceased around the beginning of May when the sheep were removed to the downlands and the meadows flooded again in preparation for the hay crop. Immediately following hay-making in July, the meadows were flooded once more, after which a second hay crop could be taken or the animals allowed to graze until Christmas.

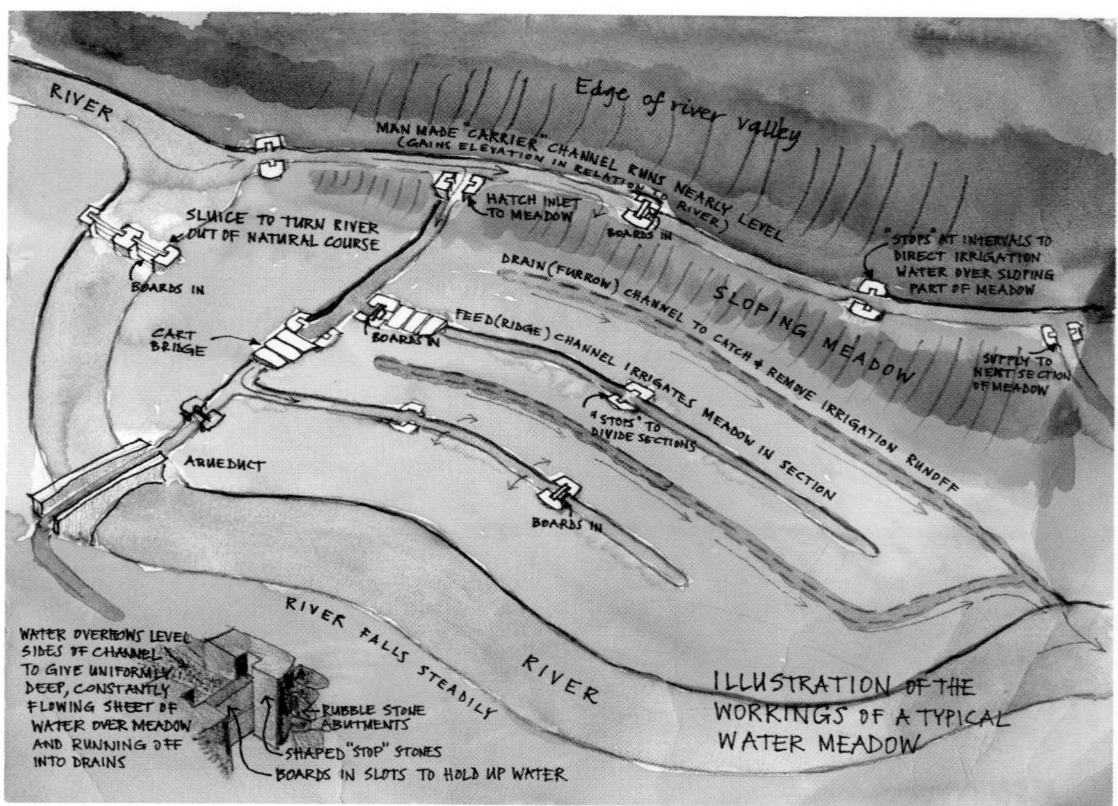

5 Diagram of a typical water meadow system — Mike Clark, Hampshire County Council

The rise of the water meadow system did not meet with universal approbation particularly from other interests sharing the river. There were fearful arguments between millers and farmers where hatches and weirs were operated to the detriment of the other party. Those enjoying the fishing rights were also affected and it is recorded on the Avon that the water meadows were blamed for serious losses in sport. Occasionally large numbers of migrating salmon smolt may have perished in the intricate system of channels if drowning coincided with a smolt run. The great Itchen angler G.E.M. Skues records having to transfer trout which had strayed during the drowning operation from runnels and drains back to the river.

Although the water meadow system helped to maintain a high soil moisture content which would have benefitted some plant and bird species, the flooding and grazing regimes coupled with the removal of all plants other than grasses, probably meant that the meadows were relatively species poor and it is now believed that they became more diverse again after their abandonment.

The decline of the water meadows

In 1852 a writer observed that 'some water meadows were not receiving as much attention as previously', and by the turn of the century the water meadows were in decline and large areas were turned over to cattle. The availability of artificial feedstuffs for sheep reduced the need for the early bite and the decline in the number of horses following the coming of the railways reduced the demand for extra hay. The biggest single factor hastening their decline was the shortage of labour particularly at the time of the 1914-18 war.

By the 1930's the water meadows were a shadow of their former selves. Drains and ditches became choked with reed and sedge. Culverts and hatches fell into disrepair and many hundreds of acres had fallen to the plough. Drainage was also affected by the lack of maintenance to the extensive system of ditches.

There can be no doubt that the water meadow system of agriculture, by its initial engineering and subsequent management and eventual abandonment, had the most radical environmental effect. It is probably the biggest single influence shaping rivers such as the Test and Itchen as they appear today. The remains of the system are still one of the most important factors in the conservation of the wildlife of the valleys and it is vitally important that ill-considered efforts to remove them, such as the ploughing in of drainage ditches, are resisted.

The development of fishing

6 Window in Silkstead Chapel, Winchester Cathedral showing Izaak Walton seated by the Itchen with St Catherine's Hill in background. *(By kind permission of the Dean and Chapter of Winchester Cathedral)* *Photo: © Dr John Crook*

There are a number of references to fishing on the chalk streams from the Middle Ages onwards. Although there are some references to angling with rod and line, it is clear that until the beginning of the 19th century the prime interest was probably as a source of food. The use of the river for fishing was secondary to the water mills and the water meadow system.

Amongst the early references to the quality of the chalk streams are Walton's comment that Hampshire 'exceeds all England for swift shallow, clear, pleasant brooks and store of Trouts' and he refers to the practice of spearing trout by the light of a torch. It is not until the beginning of the 19th century that there are any detailed accounts of the fishing. Col. Peter Hawker of Longparish kept a diary from 1802 until 1853 and the Rev. Richard Durnford of Chilbolton from 1809 until 1819. The Leckford Fishing Club kept records of the fishing from 1809, and the Houghton Club Chronicles run continuously from 1822 until this day.

These accounts show that blow lining and cross lining when there was insufficient wind to lift the blow line, as well as downstream fishing with a team of wet flies and the use of worm and minnow, were the favoured methods of fishing. Substantial catches of fish were made, particularly in May and June.

Somewhere between the beginning of the 19th century and its mid point, the use of an upstream dry fly was developed. J.W. Hills in 'A Summer on the Test' dates the invention of the floating fly to the eighteen thirties or forties. Francis Francis refers to its use on the Itchen in 1857, but it was not until the eighties with the publication of Halford's 'Floating Flies' that its use was widespread on the Test. The Houghton Club Chronicles still referred to blow lining as late as 1884.

In 1815 or 1816 the Leckford Fishing Club recorded the introduction of grayling to the Test, possibly from the Avon. Before that date they were unknown on the Test and they were no doubt then introduced into the Itchen but probably not until the twentieth century. The introduction of grayling marks the start of a still unresolved controversy about the value of grayling in a fishery and the possible adverse effects on trout populations. They have not been without both their champions and detractors. Sir Humphrey Davy asked to be an additional member of the Houghton Club solely for the grayling fishing, which Hills, who disapproved of grayling in trout streams, likened to being invited to Leicestershire for the rat hunting.

In the early years of the 19th century the fishing adapted to the river and frequently ended in June if the river was choked with weeds. As the century progressed, the river was increasingly managed both for agricultural purposes through the water meadow system, and also to maximise its fishing potential.

7 Silkstead Chapel with tomb of Izaak Walton *Photo: © Dr John Crook*

The opening of the London to Southampton railway meant that the Test and Itchen were within daily reach of eager Victorian sportsmen keen to be away from the crowds and polluted atmosphere of London. The purchase and renting of fishing, the formation of clubs and syndicates, provided funds for the improvement and maintenance of the fishing. The fishing interest also provided an incentive to maintain the rural landscape and with it the wildlife habitat. Many writers have emphasized the value placed on the beauty and interest of the river environment as being greater than the fishing interest.

With the demise of the water meadow system after the First World War, the fishing interest became the driving force in the maintenance of the river and virtually the only source of income to pay for it, apart from those owners who were willing to subsidise the costs from the agricultural incomes.

A Member of the Houghton Fishing Club.
By J. M. W. Turner, R.A.

8

The growth of regulation

Although the development of environmental regulation is normally associated with the 1970's, in practice the rivers were subject to increasing regulation by Acts of Parliament, City Corporations and later the river boards, from the beginning of the 20th century. Voluntary agreements were also established between owners through river associations such as the Test and Itchen Association formed in 1907. The records of the Association from the twenties and thirties reflect increasing concern over pollution from sewage effluent and run-off from roads. Right from its earliest days there was also some suspicion of imposed regulation. The 1940's saw the setting up of river catchment boards which were subsequently replaced by river boards. However the burden of fighting pollution cases was largely borne by owners either individually or collectively, particularly if public bodies were involved.

After the end of the Second World War, there were two further developments which were to impact adversely on chalk streams in Southern England.

9 River Test, before and after dredging *Country Life, 1948*

The first was the land drainage carried out on the lower reaches of a number of rivers including the Test and the Frome. The need to increase agricultural production was so great that the permanent adverse effects on the rivers were probably not fully appreciated at the time and the policy of land drainage continued long after the agricultural priority had receded.

The other was the growing demand for water, particularly for domestic consumption, which could be most easily met by abstraction from chalk aquifers and later direct abstraction from the rivers. Many of the schemes were fiercely opposed by owners and river associations but the legislative framework for considering objections was relatively undeveloped and the city and district councils had a legal obligation to meet all demand.

In the last twenty-five years the demand for land for residential development has also increased. Electrification of the railways and a motorway system have brought the river valleys within commuting distance of major conurbations. In consequence planning controls have been tightened to reflect the need to control such development.

The privatisation of the water industry led to the creation of the National Rivers Authority, and its successor the Environment Agency, with considerable powers over the abstraction of water, the disposal of effluent and the management of the rivers.

The developments of the past twenty-five years have raised wider concerns over the gradual erosion of the rural nature of the river valleys and the loss of their wildlife habitat. This concern has been reflected in successive pieces of legislation to protect wildlife habitats and rare and endangered species. In addition English Nature notified the Test and Itchen as Sites of Special Scientific Interest and the Itchen as a Special Area of Conservation under the European Habitat Directive.

Many owners have considerable misgivings about the extent of regulation which now faces a fishery manager, pointing to their excellent record of river management. On the other hand, they can by themselves offer little protection to the rivers from the actions of other agencies of national and local government, pursuing their own policies in such areas as road building, water abstraction and residential and industrial development. Whether we agree or not, the regulatory framework is only likely to grow and all those who are involved in the management of the rivers need to understand how it governs their day to day activities.

10 Photograph taken July 2003

10a Photograph taken April 2005

Eutrophication and low flow causing algal growth and loss of weed on the Itchen

Photos: Dennis Bright

11 The rich and varied life of a chalk stream *Photo: Michael Baron*

Chapter Two

CHALK STREAM AQUATIC PLANTS

The rich growth of aquatic plants is an obvious and distinctive feature of chalk streams. The underwater scene with its clear water, large and brilliantly green tussocks of water weeds separated by clear gravel runs is of great beauty and fascination, but the proper management of these weeds is at the very heart of a healthy river.

The term *water weed* is really a misnomer. The plants are the producers on which almost all the aquatic food chains ultimately depend and most kinds of plant have some important role to play in the ecology of the stream. Of the submerged plants, the larger are referred to as *macrophytes* and are almost all flowering plants. Many of these have emergent phases when they flower; others may flower less obviously underwater. Some of the less advanced 'lower' plants such as the mosses and algae may also form quite large masses, but smaller algae may be even more important as they are of sufficiently small size to provide food for the aquatic invertebrates. Under some conditions the growth of algae may become too great and can form scums or blanket weed which can inhibit the growth of macrophytes, clog streams and cause serious problems including toxicity of water supplies. The macrophytes will often also grow too large and they will need to be cut back; perhaps it is understandable why all these plants, in spite of their value, are still referred to as water weeds.

Conditions for growth

All these plants thrive under conditions of high light intensity, high mineral content of the water and stable, warm water; conditions described as *mesotrophic*. But as in many situations, too much may not be a good thing. Excessive concentrations of dissolved nitrate and phosphate, important components of decaying material and sewage effluent, and warm sunny conditions may promote excessive growth and the conditions become *eutrophic*. (10, 10a)

Adaptations of plants

Most aquatic macrophytes are distinguished by their long, thin and often much divided leaves that give minimum resistance to the flow of water. Many have air spaces within their stems and leaves for better buoyancy and most have either strong creeping stems or may produce a mass of roots from almost anywhere. These will help to fix the plant to the bottom and also to its neighbours as growth progresses. It will also mean aa small detached pieces root easily and allow for the spread of the plant downstream. Vegetative growth will be at its fastest from late spring to autumn, but flowering is almost always a midsummer event and is usually only evident if the weed cutting is not hard enough. Nevertheless many of the macrophytes will spread by the production of seed as well as vegetatively. Some of the weed will tend to accumulate in the quieter area nearer the bank side and this may become the first stage in a succession of growth which could eventually result in that part of the river turning into marsh or drier land.

The value of the plants

While some organic food material may come into the river from the fallen leaves of surrounding trees and shrubs, the most direct source of food for the invertebrates is the microscopic algae that grow on the stones and mud of the river bottom or attached to the larger macrophytes. The macrophytes not only provide this site but also give protection for the smaller animal life and create protected lies for fish. The patchwork development of the river bed with clumps of different kinds of weed alternating with gravel runs provides an ideal environment for a trout territory. The depth of the water will also be much influenced by the growth of weed; as summer progresses and flow volumes decline, river depth may actually increase due to the weeds impeding flow.

12 Well managed chalk stream *Photo: Michael Baron*

Problems

The proper control of weed growth is one of the most important aspects of chalk stream management. On the Test and Itchen, weed is cut according to plans prepared by the Test and Itchen Association and there are stated periods throughout the spring, summer and autumn when this is authorised. In some years, when there is a hot summer and river flows are good, weed growth may be prodigious. In other seasons cloudy water and low light intensities may result in poor growth. Whatever the situation, the weed is cut to encourage fresh growth of modest size. If left on its own it will tend to die-back at the base as the new growth shades out the old; a large clogged mass of half-dead weed may result and this may mean the end of that stretch as a fishery.

Weed growth may be poor in some seasons due to the accumulation of silt or algae on the leaves and stems. This is often associated with excessively eutrophic conditions described above, coupled with low flow velocities.

Some weeds are thought more beneficial to the river than others; Water-crowfoot *(Ranunculus)* (13) is thought to be particularly beneficial and has the added advantage that it does not accumulate too much silt and mud amongst its stems and roots. Water Starwort *(Callitriche)* (16) on the other hand has a very close growth form; there will be less epiphytic algae on its stems and so it supports less of an invertebrate population. It also accumulates mud in its roots, particularly as it is more common in areas of lower water velocity, but this provides an important habitat for water shrimps *(Gammarus)*. Control of excessive algal growth is nearly impossible; though measures to increase current can be a help.

Identification of aquatic plants

Important submerged water plants

Plants associated with fast-flowing, gravel-bottomed streams mostly in the upper reaches of the rivers.

A well-managed stream has extensive clumps of Water-crowfoot. In the south these are mostly the Brook Water-crowfoot *(Ranunculus penicillatus)* (13). The thin, dark-green leaves form large wavy clumps which may be succeeded by white flowers above the waterline in high summer. Very similar, but much less common is the River Water-dropwort *(Oenanthe fluviatilis)* (14) which is distinguished from the Water-crowfoot by its leaves being flat rather than rounded in section. This is sometimes referred to as 'carrot', though it is actually poisonous. Many stretches are heavily colonised by the bright lettuce-green masses of Water Starwort *(Callitriche)* (16) species. Although common in fast water they grow well in slower reaches and do well in the winter and also in the colder water associated with headwater springs. In some years of poor Water-crowfoot growth, Starwort may be the chief weed remaining in the river. Many high quality streams also support growths of the Lesser Water-parsnip *(Berula erecta)* (17). This is very similar to the Fool's Watercress *(Apium nodiflorum)* (18). Both have bright green leaves, but seldom form tall clumps and tend to form creeping mats with surprisingly large leaves.

Plants associated with the slower-flowing and deeper water of the middle and lower reaches of the river.

Although the species mentioned above are still found abundantly, the more varied conditions downstream support a wider range of water weeds. Deeper parts are often colonised by Mare's-tail *(Hippuris vulgaris)* (19). These are often of a brownish colour and the long, single stems are surrounded by many thin leaves giving the whole a tail-like appearance. On a much smaller scale and rather more rarely, the Water-milfoil *(Myriophyllum spicatum)* has a more divided form of growth. Quite large growths of Canadian Pondweed *(Elodea canadensis)* (20) may be found in slower-moving water and similar areas may support the larger pondweeds *Potamogeton* species.

The Horned Pondweed *(Zannichellia palustris)* (15) forms quite extensive growths distinguished by their very fine almost hair-like slimy leaves. This is easily confused with Fennel-leaved Pondweed *(Potamogeton pectinatus)*. The former has its leaves arranged opposite to one another, the latter has them alternately arranged. These are sometimes referred to as silkweed and are possibly becoming more common throughout the river system.

There are a number of rush- and reed-like species which may also colonise the river, but one of the most obvious is Common Club-rush *(Schoenoplectus lacustris)* which may have tall dark fishing-rod like stems emerging some distances above water and tape-like leaves narrowing to a pointed tip, underwater. Both fast and slower streams also support extensive colonies of Unbranched Bur-reed *(Sparganium emersum)* which have simple parallel-sided strap-like leaves with blunt round tips. These species have strong bottom-rooted rhizomes which makes their control difficult.

Everything should be done to encourage the growth of *Ranunculus*. Good water flow and a mud-free gravel substrate are the ideals that the keeper should try to achieve to help promote its growth.

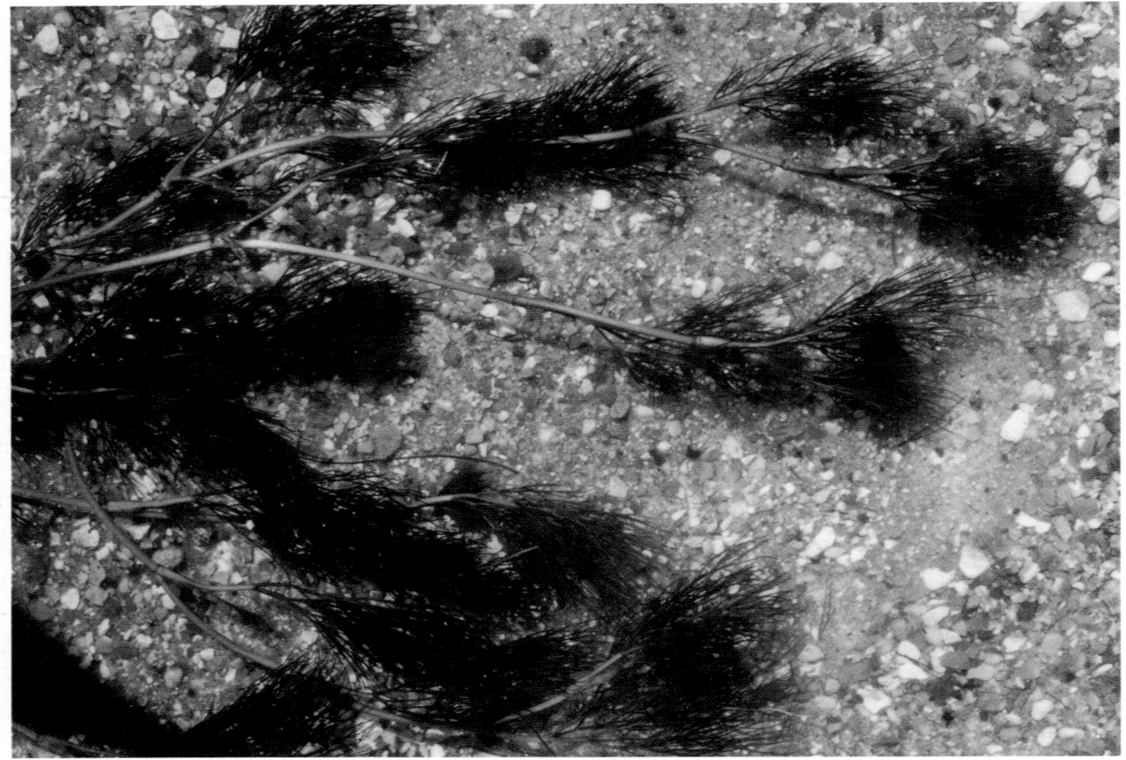

13 Water-crowfoot *Ranunculus penicillatus* *Photo: Michael Baron*

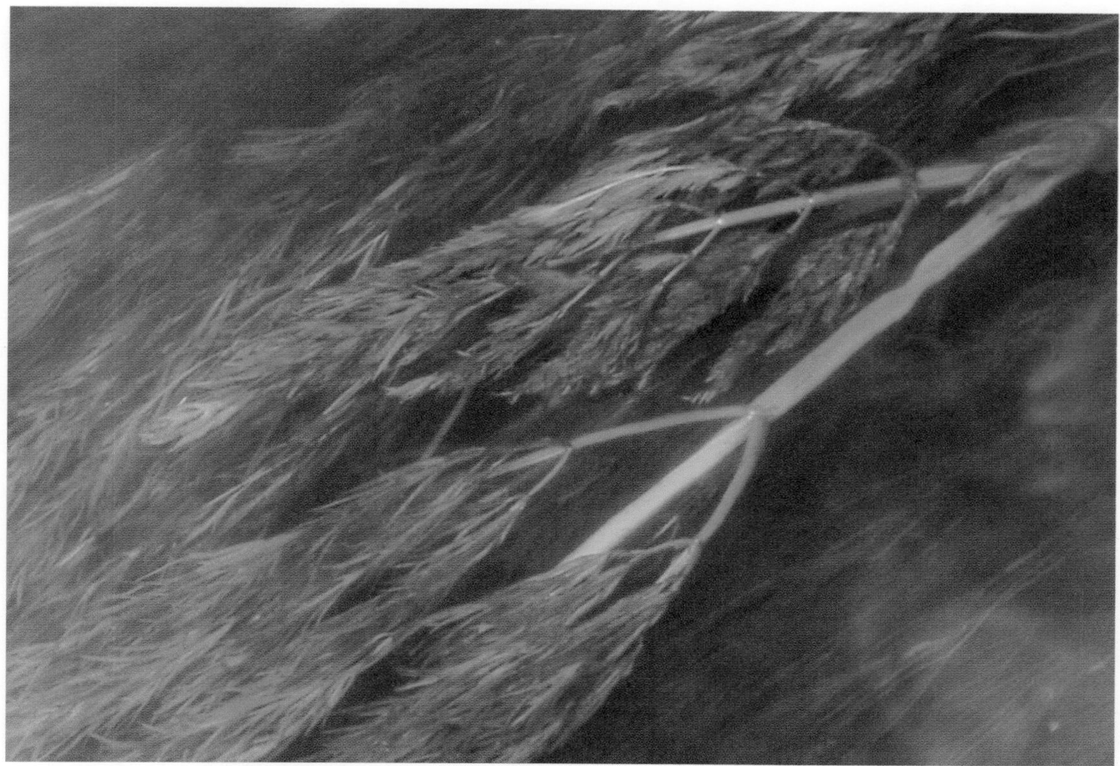

14 River Water-dropwort *Oenanthe fluviatilis* *Photo: Dr Nigel Holmes*

15 Horned Pondweed *Zannichellia palustris* *Photo: Dr Nigel Holmes*

16 Water-starwort *Callitriche* sp. *Photo: Michael Baron*

17 Lesser Water-parsnip *Berula erecta* *Photo: Dr Nigel Holmes*

18 Fool's Watercress *Apium nodiflorum* *Photo: Dr Nigel Holmes*

19 Mare's-tail *Hippuris vulgaris* *Photo: Michael Baron*

20 Canadian Pondweed *Elodea canadensis* *Photo: Michael Baron*

21 Willow Moss *Fontinalis antipyretica* *Photo: Michael Baron*

Table of submerged, emergent & floating aquatic plants commonly found in chalk streams

The larger and more abundant plants are listed first.

Common Name	Latin Name	Leaf Features	Flowers	Distribution
Water-crowfoot (13)	*Ranunculus* species	Much divided, rounded in section. Forming large clumps	White, above water	Abundant
Water-starwort (16)	*Callitriche* species	Small lettuce green leaves forming a large bright green clump. Underwater leaves thin, floating leaves often spoon-shaped	Below water	Abundant
Lesser Water-parsnip or Water-celery (17)	*Berula erecta*	Dull green with 7-10 segments. Pale ring on leaf stalk	Above water	Very common, particularly in the main stream
Fool's Watercress (18)	*Apium nodiflorum*	Bright, shiny green, with 2-6 segments. No pale ring on leaf stalk.	Above water	Common, particularly towards river margin.
Mare's-tail (19)	*Hippuris vulgaris*	Stems long, often over 1m. Unbranched. c10 leaves arranged in whorls each leaf c30x2mm	Spikes above water	Common in deep water
River Water-dropwort (14)	*Oenanthe fluviatilis*	Much divided, very thin, flattened in section. Forming large clumps	In umbels above water	Common Poisonous
Common Club-rush	*Schoenoplectus lacustris*	Thin and tape like underwater. Long 'fishing rods' above	Above water	Occasional, particularly in deep water

Common Name	Latin Name	Leaf Features	Flowers	Distribution
Horned Pondweed (15)	*Zannichellia palustris*	Long and silky, opposite	Below water	Occasional
Canadian Pondweed (20)	*Elodea canadensis*	c3 leaves arranged in whorls. Each leaf <10mm long	Seldom flowers	Common in muddy streams (non-active)
Willow Moss (21)	*Fontinalis antipyretica*	Small(c5mm) dark green. Forms clumps	None	Occasional on stones and gravel
Water Milfoil	*Myriophyllum spicatum*	Much divided to give fern-like mass. Often slightly reddish colour	Inconspicuous	Common in lower reaches
Opposite-leaved Pondweed	*Groenlandia densa*	3cm x10mm arranged opposite to one another to form a flattened 2 sided shoot	Inconspicuous	Occasional
Common Duckweed	*Lemna minor*	Very small rounded floating leaves	Inconspicuous	Abundant
Ivy Duckweed	*Lemna trisulca*	Very small, ivy shaped leaves. Submerged	Inconspicuous	Common
Curled Pondweed	*Potamogeton crispus*	Wavy 9 cm x 1cm. Forms large clumps	Inconspicuous	Occasional, in lower and deeper reaches
Fennel-leaved Pondweed	*Potamogeton pectinatus*	Long and silky, alternate	Below water	Occasional, nearer river margin
Broad-leaved Pondweed	*Potamogeton natans*	Submerged leaves up to 40cm long. Floating leaves up to 60cm long	Inconspicuous	Occasional, in deep and sluggish water
Unbranched Bur-reed	*Sparganium emersum*	Thin, tape like, but triangular in section	Above water	Occasional
Azolla Fern (22)	*Azolla filiculoides*	Very small bluish-green fronds turning reddish in Autumn	None	Increasingly in still water and in river margins (non active)

The algae

The variety of algae

It is really only under eutrophic, bright and warm conditions that the presence of algae becomes noticeable. Long strands or skeins of blanket weed *(Cladophora)* (23) may develop, the bottom may become coated in greenish slime and the macrophytes become covered with a brown-green scum. Even the flowing river may become green due to free-floating algae. Algae attached to other plants are described as *epiphytic*, those attached to stones and the bottom as *epilithic;* and those that are free-living are *planktonic*. Even when there are no algae visible at a cursory look, examination of a macrophyte leaf or scraping a stone will reveal an abundance of algae when the material is examined under the microscope.

The main kinds of algae found in chalk streams:

a) Multifilament forms

Batrachospermum forms a small reddish tussock, epilithic, in fast water. Uncommon.

b) Filamentous forms, sometimes branched

Cladophora (23): This forms small and large skeins of a deep green colour. These are sometimes several metres long. The filament is composed of separate cells. Abundant. *Vaucheria*: This forms short bright green tussocks. The filaments are tube-like. It often develops on mud. Abundant.

c) Desmid forms. Microscopic crescent-shaped cells, found sporadically throughout the rivers.

d) Diatoms. These microscopic forms occur in a multitude of shapes. All have a silica-containing wall to their cells. Most of the forms are very common indeed, both epilithic and epiphytic.

e) Blue-green algae. These often form an almost black scum-like growth.

Seasonal changes in algal growth

In many lake and river systems, algae show a gradual increase in abundance during the spring as the water becomes warmer. Later there is often a decline as the dissolved minerals necessary for growth become scarce due to their absorption by both the algae themselves and the development of the stronger growing water weeds. Late in the year as the growth of the macrophytes begins to decline with the onset of autumn some species of algae may show a second peak of growth due to the release of further dissolved minerals as the decay of autumnal leaves begins. In the chalk streams these trends may occur in some microscopic and planktonic forms, but here mineral availability is less likely to limit plant growth, and the growth of the more obvious algae such as the blanket weed *(Cladophora)* may show a continued increase during the summer months, only dying back in late autumn and early winter. Under some conditions, a mucilaginous mass of dead and alive algae, bacteria and silt, may rise to the surface and form unpleasant floating rafts.

There seems little doubt that high levels of nitrate and phosphate coupled with reduced water flows, high temperatures and high light intensity all combine to stimulate the growth of *Cladophora*. This plant can grow alarmingly fast, forming skeins many metres long, shading out the macrophytes and generally producing an unsightly and unwelcome mass. This growth is unpopular with fishermen as their lines get caught and as the alga begins to decay, oxygen-deficient (anaerobic) and unwholesome conditions may result. Planktonic blooms of algae have been known to occur in the lower reaches of the river and these free-living algae may contribute to the turbidity of the stream. They would normally only be expected in high summer when flows are reduced and the concentration of dissolved minerals becomes unusually high.

Control of algal growth

Unfortunately there is comparatively little that the keeper can do to control excessive growths of algae. They grow fast and even if the gravels are frequently raked clean, they are likely to recolonise the area within days. Cutting out of the larger skeins of *Cladophora* is difficult enough as they tend to slip through the rake or scythe. The only control possible is to keep the river flowing fast over a good growth of *Ranunculus* (13) and the other macrophytes so as to reduce the chances of algae gaining a hold. Macrophyte growth may reduce concentrations of some minerals such as nitrates and algal growth may be less severe. Weed booms may help to skim off floating algae, and in some smaller streams straw bales have proved effective in reducing algal bloom.

Everything possible must be done to ensure that the water is of good quality and that sewage effluents are scrubbed as much as possible of the minerals that encourage algal development.

Legislation to reduce abstraction during the summer months must also help to alleviate the problem. Land management and farming practice in lands adjacent to the head-waters and higher parts of the river catchment may play an important part in determining the concentrations of dissolved nutrients. Run-off following fertilizer treatment, especially in springtime, may result in unwarrantably high levels of nutrients finding their way into the river.

There is a case for encouraging the development of woodlands, permanent pastures and unfertilised areas around the headwaters and beside streams. Ploughing and intensive management close to the rivers is to be discouraged. Promotion of schemes such as Countryside Stewardship and Environmentally Sensitive Areas will also help to reduce the effects of some agricultural practices, near headwaters and watercourses generally.

If we are in for a period of increasingly hot summers, then the spread of algae in our chalk streams is likely to become more than a nuisance and turn into a serious problem.

Management of aquatic plants

There is no real knowledge of when the river weed was first managed. No doubt with the advent of water mills, fulling mills, navigation and drainage schemes it became necessary for some kind of river management to take place and for the streams to be kept clear of obstruction. Although some management may have taken place in Roman times and even before, there can be little doubt that little organised weed management for reasons

22 Azolla Fern *Azolla filiculoides* *Photo: Michael Baron*

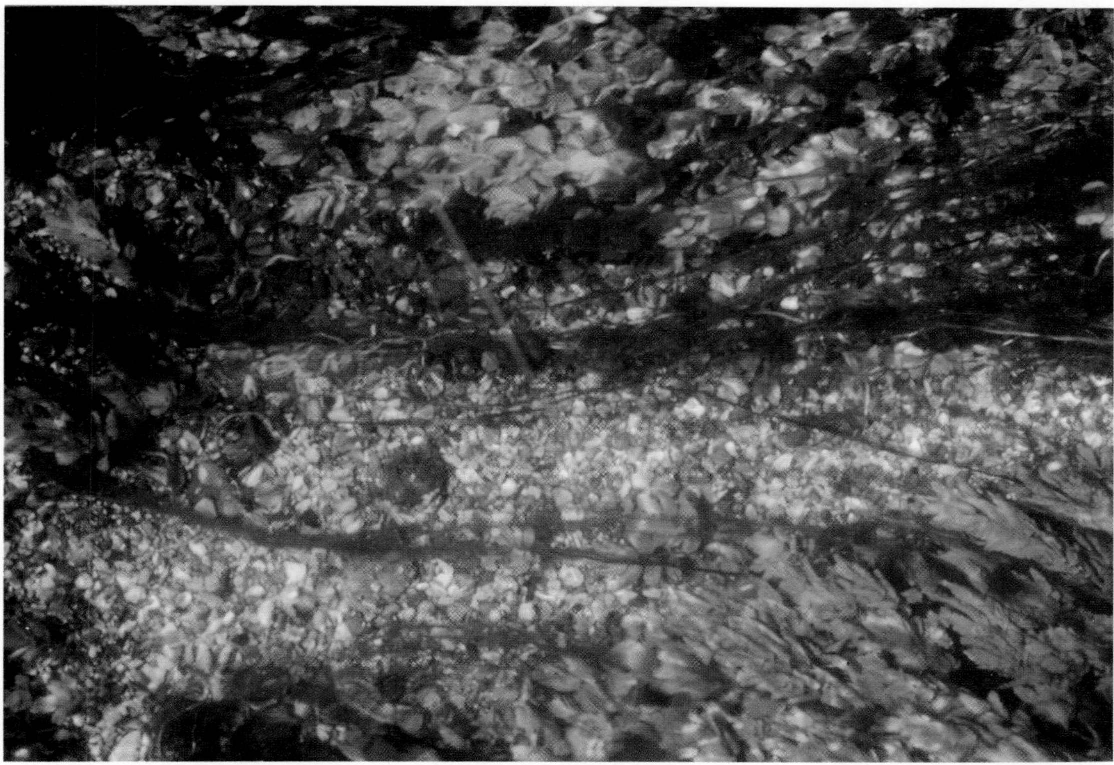

23 Long skeins of algae *Cladophora* *Photo: Michael Baron*

of fishery management was carried out prior to the nineteenth century. In the Reverend Durnford's day on the Test at Chilbolton early in the nineteenth century, fishing ceased by the end of June when the river became choked with weed and unfishable. Later with the advent of river keepers, one of their main tasks was the proper cutting of weed. In the early days this may have been a disorganised activity making the river unfishable for long periods when mud and cut weed would be floating downstream. Even now with a carefully planned and staggerred weed cutting programme and efforts to remove as much weed as possible near to the site of the cutting, the presence of cut-weed floating downstream and the frequent build-up of masses of floating weed around the banks and bridge supports can pose real problems for both keepers and fishermen. In more recent years weed growth in many parts of the Test and Itchen has been poor as a result of excessive algal growth. However it remains an important activity for flood risk management, and management of the river habitat generally as well as for fishery management purposes.

Growth and control of aquatic plants

Every season produces its own problems on the river and this is never more true than when examining weed growth. Some seasons produce little growth, at other times some of the stronger growing such as Water-crowfoot *(Ranunculus)* (1) may grow at a great rate and tax the keeper's skills and resources. The aim should be to start the season with a good gravel bottom with distinct, separate and moderate sized clumps of weeds. They should grow rapidly as the water temperature and light intensity increase and will normally be subject to a 'trimming' cut in April.

In some years, perhaps due to excessive silt, algal growth or very cold water conditions, there may be little growth and no need for a serious cut. One thing is certain and that is that any weed is better than no weed, so if weed growth is poor it may be necessary to plant additional *Ranunculus* from reaches where it has managed to survive in a healthy state. Transfer of weed should only be from parts of the same river system.

Much concern has been expressed about the effect of swan grazing on *Ranunculus*, particularly when river levels are low and growth poor. The main problem appears to be with flocks of immature birds, which the normal resident pair are unable to drive away.

The total clearance of weed over a large stretch is almost certain to result in a decline of invertebrate life and a reduction in the number of fish.

Value of different aquatic plants

Some plants are generally accepted as being more useful in the river than others and while it is generally wise to encourage as great a diversity of species as possible, some plants do support a larger and possibly more varied population of invertebrates than others. Some also provide better cover for trout, which can lie under or below the weed tussock; other kinds are not excessive accumulators of mud and silt. The most favoured are Water-crowfoot *(Ranunculus)* (13); Water-celery *(Berula erecta)* and the similar Fool's Cress *(Apium nodiflorum)* (18) .

Less popular with keepers is the Mare's-tail *(Hippuris vulgaris)* (19), perhaps because it does not look an attractive species and is found in slower and more muddy reaches, nevertheless it supports a rich fauna of nymphs of Ephemeroptera. All these four species are relatively easily managed and can be cut easily with a scythe. Of the other common weeds, Starwort *(Callitriche* species) (16) does hold a reasonable population of animals, particularly freshwater shrimps *Gammarus*. However, it forms such a thick network of stems and leaves that it acts as a filter and in consequence accumulates much mud and silt. For this reason it was not always very popular in streams and was often cut (or raked) particularly hard.

Finally it must be emphasised that all healthy green weeds are oxygenators and that some of the common weeds, particularly Water-crowfoot, Water-dropwort, Water-celery and Water-parsnip are particularly important in this way. Indeed on a bright summer's day, the oxygen saturation of the water may rise well over one hundred percent, becoming supersaturated. For this reason, if no other, it is essential to encourage a good growth of healthy weed.

It should be pointed out that all green weed, as well as decaying matter, will be utilising oxygen during the night time and this, even in a healthy river, will result in a depletion of oxygen during the dark. Oxygen levels also depend on temperature: the higher the temperature the less oxygen the river will hold. Fish metabolic rates will also be higher and their oxygen consumption will increase at these higher temperatures and so warm conditions at night may, in extreme conditions, pose a special problem. In these extremes, riffles and waterfalls, by allowing for a greater exchange of gases, may be important in maintaining adequate oxygen saturations.

Aquatic and marginal vegetation

Emergent vegetation

Many of the truly aquatic plants have leaves and flowering stems which emerge from the river in high summer if weed cutting is not severe. The white flowers of Water-crowfoot *(Ranunculus)* (13) are an obvious feature of many rivers. Less well-known are the large flowering stalks of Water-dropwort *(Oenanthe fluviatilis)* and the almost conifer like flowering-stems of the Mare's-tail *(Hippuris vulgaris)* (19). Many of these plants and also Starwort *(Callitriche)* (16) have leaves which differ greatly from those underwater. While most keepers would prefer to keep these plants trimmed so that they do not emerge, in many instances weed rafts develop and flowering stems emerge. However this emergent vegetation is useful in allowing foothold for many aquatic insects when they crawl out of the water to moult into their flying stage. They also provide nesting sites for aquatic birds such as coot and dabchick.

The river margin

Due to the slower current several important plants occur close to the bank and form part of the bank side vegetation. This is a varied habitat and the species present will depend on the nature of the bank; whether it is gravel, chalk, peat or mud, and the depth the of water. Common marginal plants include Watercress *(Rorippa nasturtium-aquaticum)* (24), Water Forget-me-not *(Myosotis scorpioides)* (26), Water-speedwell *(Veronica anagallis-aquatica)* (25), Water-mint *(Mentha aquatica)* (27), Hairy Willow-herb *(Epilobium hirsutum)* (28) and Purple-loosestrife *(Lythrum salicaria)* (29) as well as the alien introductions, the balsams *(Impatiens)* (30) and Monkeyflower *(Mimulus guttatus)* (31). All these are attractive species and add greatly to the diversity and interest of the river margin.

24 Watercress *Rorippa nasturtium-aquaticum* *Photo: Michael Baron*

25 Water-speedwell *Veronica anagallis-aquatica* *Photo: Warren Gilchrist*

26 Water Forget-me-not *Myosotis scorpioides* *Photo: Warren Gilchrist*

27 Water Mint *Mentha aquatica* *Photo: Warren Gilchrist*

28 Hairy Willowherb *Epilobium hirsutum*
Photo: Michael Baron

29 Purple-loosestrife *Lythrum salicaria*
Photo: Michael Baron

30 Orange Balsam *Impatiens capensis* (non-native)
Photo: Michael Baron

31 Monkeyflower *Mimulus guttatus* (non-native)
Photo: Michael Baron

32 Bur-reed *Sparganium erectum*
Photo: Dr Nigel Holmes

33 Greater Tussock-sedge *(Carex paniculata)*
Photo: Michael Baron

33a Common Reed *(Phragmites australis)*
Photo: Michael Baron

There are also a number of grasses, reeds, sedges and rushes that are usually more common where the current is slower and a more muddy margin develops. Float Grass *(Glyceria fluitans)* is quite commonly found especially on drowned water meadows. Reeds can be common, especially the tall Common Reed *(Phragmites australis)* (33a) which may still be cut for thatching from the older and more extensive reed beds. Winter cutting of *Phragmites* may be beneficial to the survival and health of the plant in the subsequent year. Bur-reed *(Sparganium erectum)* (32) can form large stands at the water's edge, but Great Reedmace *(Typha latifolia)* although widespread, is rather less common. Sedges, distinguished by a V-shaped channel along the centre of their leaves, can form extensive marginal groups. There are many species, but the most distinctive is the Greater Tussock-sedge *(Carex paniculata)* (33) which forms aged almost palm-like tussocks both at the river margin and on drier ground. Some of these tussocks may be more than fifty years old and they also provide a home for a number of interesting invertebrates including, occasionally, the rare Desmoulin's Snail.

Some of these reeds and sedges form useful marginal cover that screens the river from paths along the bank. Many grow too strongly and need cutting back, but they may help to consolidate an otherwise unstable and marshy area. They can also help to reduce the width of a stream which has become overwide.

34 Barn Owl quartering the water meadow *Photo: Dennis Bright*

Chapter Three

THE ECOLOGY OF THE RIVER BANK

The riverene and bank side habitat

Importance

Part of the enjoyment of fishing chalk streams such as the Test or Itchen undoubtedly lies in their surroundings. The moist soil of the surrounding fen land supports an incredibly rich and varied series of habitats from reed beds to marsh, from willow scrub to quite well-developed oak wood. The trees and bushes provide useful screening from houses, roads and other unsightly areas, they act as windbreaks, reducing the chances that fly life will be blown off the river and also provide roosts and nesting sites for numerous species of birds. The contrast between a reach where there are sizeable surroundings of this kind and an insensitively managed stretch where there is a canal-like river bounded by a barbed-wire fence and a few poplar trees is only too apparent. The river is, of course, a continuous system with a pattern that varies not only with the seasons but from year to year. Nothing is static and the job of the river keeper is to try to maintain this continuous, variable riverine corridor, not only for the migration of fish but for the various animals and migratory birds that move up and down the river system.

Range of habitats

The different habitats relate directly to the soil water content; ranging from the truly aquatic to the dry terrestrial. This gradual change or *cline* in soil moisture produces a series of small-scale habitats: the river margin, the river bank, marsh and fen land, grassland and even scrub and woodland. Given time, the more open communities may become colonised by shrubs and trees and a succession of changes may take place which could result in the formation of mature woodland. Management may arrest these changes; grass may be cut, grazing initiated, scrub removed and trees cut down.

There may be former carriers, now largely silted up, old ponds and reed beds. These will gradually dry up and conscious management decisions may have to be made as to their future. If the area is extensive then economic considerations alone will dictate that only a small part of the site is cut, cleared or specially managed at a time. This 'patchwork' system of management means that only a relatively small part is disturbed at one time and that wildlife may be able to move from one part of the site to another and so the value of the whole is maintained.

Vegetation of the riverside

It is surprising how quickly the river margin accumulates silt, becomes colonised and turns into a small marsh. The bank can be amazingly rich. It is good to see a wealth of attractive flowers and a screen between the fisherman and the river is desirable. Much of the vegetation can be cut to about 0.5m which may still allow flowering to take place. There will be a number of less welcome species such as nettle and hogweed and these can be selectively cut; in any case it is likely that these will be abundant under trees further from the bank.

Plants of the riverbank

The following table lists a number of the more attractive species found on the river bank. The species which are usually more common are listed first.

Marsh Marigold *Caltha palustris* (36)	Bright yellow in spring	Common alongside ditches and marshy meadows
Hairy Willowherb *Epilobium hirsutum* (28)	Tall, pink flowers in summer	Common on the bank side
Purple-loosestrife *Lythrum salicaria* (29)	Tall, purple flowers; summer	Common on the bank side
Comfrey *Symphytum officinale* (37, 38a,b)	Large rough leaves, hanging tubular flowers white to purple. Food for the scarlet tiger moth caterpillar	Very common in meadows and rough areas besides the river
Flag Iris *Iris pseudacorus* (39)	Bright yellow flowers	Common around marshy areas, ponds and the bank side
Woody Nightshade *Solanum dulcamara*	A woody scrambling plant. Purple flowers and red fruits	Common on the bank side
Meadow-sweet *Filipendula ulmaria* (40)	White heads of small flowers	Common on the drier bank sides
Water Dock *Rumex hydrolapathum* (42)	Tall pointed leaves	Common on the bank side
Water-speedwell *Veronica anagallis-aquatica* (25)	Flowers pink-blue	Common on the wet river margins and as a mud coloniser.
Brooklime *Veronica beccabunga*	Bright blue flowers	Common in marshy areas
Water Forget-me-not *Myosotis scorpioides* (26)	Bright blue flowers	Common on the wet river margin
Water Mint *Mentha aquatica* (27)	Pale lilac flowers in summer. Strongly aromatic smell	Common on the bank side
Marsh Woundwort *Stachys palustris* (41)	Atractive pale pink-purple flowers	Occasional on the drier bank sides
Fleabane *Pulicaria dysenterica* (43)	Yellow daisy flowers. Hairy leaves	Common on the bank side

Hemp Agrimony *Eupatorium cannabinum* (44)	Tall, with heads of pale purple-white flowers	Common on the bank side
Monkeyflower *Mimulus guttatus* (31)	Bright yellow flowers in summer with red spots Explosive seeds. Alien	Common at the bank side
Orange Balsam *Impatiens capensis* (30)	Orange flowers in summer, explosive seeds. Alien	Common on the bank side
Yellow Loosestrife *Lysimachia vulgaris*	Tall, yellow flowers; summer	Occasional on the bank side
Water Avens *Geum rivale* (45)	Pink-orange hanging flowers	Quite common in the wet meadows of the southern chalk streams. Rarer elsewhere
Hemlock Water-dropwort *Oenanthe crocata* (46)	Tall, with white umbells. Much divided leaves. Very poisonous	Occasional on the bank side
Southern Marsh Orchid *Dactylorhiza praetermissa* (47)	Deep purple flowers. No spots on leaves	Quite common in wet meadows where the grass is short
Common Spotted Orchid *Dactylorhiza fuchsii*	Spotted leaves	Quite common in the rather drier meadows and on chalk made-up banks
Meadow Orchid *Dactylorhiza incarnata*	Rarer than above, flowers salmon pink. Early flowering	Rare, in wet meadows
Hybrid orchids	Many hybrids between the orchids are often found	
Sweet Flag *Acorus calamus*	Like the flag iris but with curious green arum-like flower, leaves very sweet smelling	Rare. Usually around the margins of still water
Common Meadow-rue *Thalictrum flavum* (49)	Tall. Small white flowers. Leaves with many leaflets	Rare. Scrubby areas beside the river
Common Figwort *Scrophularia nodosa* (48)	A tall, strong plant. Stem square in section. Small snapdragon flowers	Quite common on the bank side
Ragged-Robin *Lychnis flos-cuculi* (51)	Bright pink flowers	Local on older grassy meadows

35 Reed Sweet-grass *Glyceria maxima*
Photo: Michael Baron

36 Marsh Marigold *Caltha palustris*
Photo: Michael Baron

37 Comfrey *Symphytum officinale* with Hogweed in foreground *Photo: Warren Gilchrist*

38a, b White and Purple Comfrey *Symphytum officinale* *Photos: Michael Baron*

39 Flag Iris *Iris pseudacorus* *Photo: Michael Baron*

40 Meadow-sweet *Filipendula ulmaria*
Photo: Michael Baron

41 Marsh Woundwort *Stachys palustris*
Photo: Michael Baron

42 Water Dock *Rumex hydrolapathum* *Photo: Warren Gilchrist*

43 Fleabane *Pulicaria dysenterica* *Photo: Michael Baron*

44 Hemp Agrimony *Eupatorium cannabinum*
Photo: Michael Baron

45 Water Avens *Geum rivale* amongst Reed Sweet-grass
Photo: Warren Gilchrist

46a,b Hemlock Water-dropwort *Oenanthe crocata* *Photo: Warren Gilchrist*

47 Marsh Orchid *Dactylorhiza praetermissa*
Photo: Michael Baron

48 Figwort *Scrophularia nodosa* *Photo: Warren Gilchrist*

49 Common Meadow-rue *Thalictrum flavum* *Photo: Warren Gilchrist*

50 Yellow Loosestrife *Lysimachia vulgaris*
Photo: Michael Baron

51 Ragged-Robin *Lychnis flos-cuculi*
Photo: Michael Baron

Alien species

Some of the more attractive species found along the banks are aliens that have spread along the rivers during the last fifty or so years. The Monkeyflower *(Mimulus guttatus)* (31) is native of western north America, while the more recently introduced Orange Balsam *(Impatiens capensis)* (30) is from the eastern part of America. A potentially more serious invader is the Himalayan Balsam *(Impatiens glandulifera)* (52). This grows to about 2m and although common in the north is only recently beginning to pose a problem in the south. Two of the most problematic introductions are Japanese Knotweed, which is particularly difficult to eradicate other than by spraying, and Giant Hogweed *(Heracleum mantegazzianum)* which forms tall and dominant plants which shade out the native vegetation and can also induce serious rashes and poisoning.

52 Himalayan Balsam *Impatiens glandulifera*
Photo:Lawrence Talks

Warm weather and low flows may aid the growth of the small floating fern *(Azolla filiculoides)*. This has small reddish leaves which grow rather like duckweed. It can become a pest in enclosed waters and some streams.

Although these introductions often add an attractive element to the riverside they may oust local species and eventually cause management headaches. The plants may have originated in gardens alongside the river and escapes from these should be watched very carefully.

Sedge and reed beds

The tall thin stems of many of the grass, reed and sedge families are a characteristic feature of most riverine communities. Some may be of annoyance to fishermen, but they provide good cover for birds such as reed and sedge warblers and are important colonisers of old carriers and wet fenland. In a few instances the Common Reed *(Phragmites australis)* grows so well that it may be cut on a commercial, rotational basis for thatch, but smaller patches are of little use and occasional cutting and even burning may be recommended (after taking proper advice and due precautions). All the rushes and reeds add considerably to the attractiveness of the area and apart from holding large colonies of invertebrate animals, often consolidate areas that might otherwise be subject to erosion. It has also been shown that such beds may form very useful filters, removing potentially damaging effluent from the watercourse.

The variability and patchwork nature of the surrounding plant communities add immeasurably to the natural history of the area; where one area is subject to natural ageing, damage or over-management, then another site may act as a refuge. This pattern will be subject to continuous change over the years.

Table: The more common reeds, rushes and sedges of the riverine Area

Common Reed *Phragmites australis* (33a)	Leaves bluish-green, sharply pointed, bending in the direction of the wind. Large spreading flower heads	Forms extensive reed beds
Reed Canary Grass *Phalaris arundinacea*	Leaves bluish green, sharply pointed. Smaller flower heads in clumped spikes	A common tall, tough grass, like a smaller version of the Common Reed.
Great Reedmace *Typha latifolia*	Flowers form a thick, dark brown, club-like spike	The large bulrush of still water
Bur-reed *Sparganium erectum*	Triangular sectioned stems Flowers develop into greenish burrs	A tall plant, common along wet bank sides
Reed Sweet-grass *Glyceria maxima* (35)	Leaves bright green with boat shaped tips. Large pale flower spikes	A strong, common grass.
Greater Tussock-sedge *Carex paniculata* (33)	Triangular sectioned stems. Leaves are keeled and saw edged	Forms very large tussocks.

Lesser Pond-sedge *Carex acutiformis*	The stems are triangular in section. Tough, greyish leaves are keeled	Forms large continuous bank side colonies
Common Club-rush *Schoenoplectus lacustris*	Long, dark green, fishing-rod like stems. Tape-like underwater leaves	Occasional, in deep water and slow stretches

Watercress

Apart from their presence in commercial cress beds, wild or escaped plants of Watercress *(Rorippa nasturtium-aquaticum)* may form a conspicuous feature of the river margin and overgrown carriers. In the river they are a good site for invertebrates, particularly snails and shrimps, and are a favoured site for small fish and fry. In high summer a marginal growth of cress often provides a lie for larger trout.

Commercial cress beds are of considerable economic importance, the plants growing particularly well in the clear water of spring-fed streams, the temperature of which remains remarkably constant, remaining relatively warm in winter and cool in summer. Some of these beds may be augmented by boreholes and this may help maintain river flows in times of drought. However, in spite of settling beds and careful management, some silt often escapes into the river when the beds are cleared out and in addition there may be problems through the use of fertilisers, basic slag, zinc and pesticide treatments. Occasional serious pollution incidents have also occurred when accidental spillages of the materials used for washing the cress have escaped into the rivers. The unauthorised collection of watercress for domestic use is to be discouraged as it may be growing in poorer quality water or where there is a possibility of infection from liver fluke, a dangerous parasite that has hosts in water snails, sheep and even man.

Cut paths and grassland

Fishermen usually require easy access to the river bank and most bank sides are cut or mown several times a year. This encourages the shorter common meadow grasses and an interesting flora may develop which will be different to most other communities in the area. Here the smaller and more vulnerable species may be able to compete successfully. Buttercups, clover, Cuckooflower, also known as Lady's Smock *(Cardamine pratensis)*, and orchids may appear and increase. Avoidance of spraying and the absence of fertilisers will encourage these species. Sensitive and selective cutting will also help; avoidance of flowering orchid colonies will enable them to seed and the colony may spread surprisingly fast even though their seeds may take over five years to reach flowering.

The paths close by the bank are normally cut about once a month; less important areas, often those further back from the bank, need to be cut no more than two or three times a year. It is generally beneficial to give all paths a good cut late in the season, perhaps in September, so that grass tussocks do not develop and the smaller herbs get a good start in the spring.

Heavy cutting machinery and excessive trampling can break up the turf and in such areas slatted planking may have to be used to avoid damage.

Fenland meadows

Traces of old grazed meadows may still be seen close to the river bank and may hold an amazingly rich and varied flora. Land around some of the old headwater springs supports some of Hampshire's rarest plants and their continued survival may depend on the maintenance of a grazing regime. Sometimes cattle have access to the river. This may cause problems through disturbance and mudding, particularly if they eat *Ranunculus*. Unlike most buttercups this is a palatable species and in hot weather cattle may feed quite extensively on it. Although their treading of the meadow may create microhabitats the benefits of this are usually outweighed by the loss of marginal vegetation and the introduction of mud into the river.

Carr woodland

Fen and marsh, if left alone, will gradually develop through a *succession* into mature woodland. The normal sequence involves colonisation by scrub willows, ash and alder and finally oak. These woodland communities on base-rich soil are referred to as *carr woodland*. Many bank side areas show examples of this happening, albeit on a small scale, but their importance is considerable. The trees consolidate the ground, provide screening, cover and habitats for a vast range of small insects and birds. Beyond all else they add immeasurably to the appearance of the area. Sometimes tree species that have special commercial value, such as Cricket Bat Willows and poplars may be planted. In the past many of the plantations could be criticised for being too regimented. Now the emphasis should be more on the planting of native species in informal groups. Conifers, spruce, pine and cypress which are not native to the area seldom thrive on these soils and are best avoided altogether.

Table of native trees and larger shrubs suitable for planting adjacent to river banks.

Oak	*Quercus robur*	Slow growing; grow surprisingly well on wet and rich soils, but seldom form timber-quality trees
Alder	*Alnus glutinosa*	Usually fairly small trees which grow well on these soils. Can be pollarded
Taller willows	*Salix alba etc*	Attractive, fast growing, but older plants tend to break up and may cause problems if they fall in the river
White Poplar	*Populus alba*	A tall tree with attractive coloured leaves. Best planted well back from the bank. Probably introduced
Shrubby willows	*Salix caprea etc*	Useful as screening plants, should not be planted close to the riverside

Planting should not be so close to the river as to cause undue shading, but the occasional tree overhanging the river makes for variety and produces a feature of special interest and indeed often turns out to be the lie of a good fish. In general, although economics may suggest that the river should be edged perhaps twenty metres from the bank side with a line of trees, there may be much to be said for planting occasional blocks at right angles to the line of the river. These will be particularly effective in reducing wind, which usually tends to blow up or down the river valley. In all these plantings it is much better to stick to native species and to avoid the use of aliens.

Finally, it must be emphasised that although the above concentrates on the great variety of the plant life in the fenland, the plants will support a multitude of small invertebrate animals which may be nearly as important to the fish as those animals living within the water.

Animal life of the riverside

Insects

The vegetation surrounding the river supports a number of conspicuous and attractive species. Butterflies are often common; nettles provide food for Tortoiseshell, Peacock and Red Admirals. Speckled Wood and Meadow Brown feed on grasses, Small White on Water Cress, Orange Tip on Lady's Smock and Brimstone on Buckthorn. The day-flying Scarlet Tiger moth is one of the most spectacular insects of the fen and feeds especially on Comfrey as well as other plants. The importance of the bank side vegetation as foods for the larger moths and butterflies is illustrated by the table below.

Table showing the relative importance of some of the more common trees, shrubs and herbs of the bank side as food plants for the larger butterflies and moths

Over 100 species... A
50-100 species... B
10-50 species... C
1-10 species... D

TREES		
Oak	*Quercus robur*	A
Birch	*Betula pendula*	A
Goat Willow	*Salix caprea*	A
Hawthorn	*Crataegus monogyna*	B
Blackthorn	*Prunus spinosa*	B
Alder	*Alnus glutinosa*	B
Ash	*Fraxinus excelsior*	C
Poplar	*Populus spp.*	C
SHRUBS		
Willows	*Salix spp.*	B
Dog Rose	*Rosa canina*	C
Bramble	*Rubus spp.*	C

HERBS		
Grasses	*Poa, Festuca, etc*	A
Docks	*Rumex spp.*	B
Reed	*Phragmites australis*	C
Nettle	*Urtica dioica*	C
Primrose	*Primula vulgaris*	C
Dandelion	*Taraxacum officinale*	C
Hairy Willowherb	*Epilobium hirsutum*	D
Purple-loosestrife	*Lythrum salicaria*	D
Comfrey	*Symphytum officinale*	D
Hemp Agrimony	*Eupatorium cannabinum*	D

Many of the riverine insects will 'lay-up' in the trees and shrubs of the river margin. In exposed areas all these are much more likely to be blown away from the river. Indeed this has often been suggested as the reason for the long demise of the mayfly from some reaches of the river following a series of years with strong easterly winds.

Birds

The bird life of the river can be divided into three groups: those that feed largely on aquatic plants, those that eat fish, large and small, and those that eat seeds and insects beside the watercourse. The main herbivores are swans, ducks, Canada Geese, Coot and Moorhen, though all of these will probably also feed on floating insects. The smaller Coot and Moorhen are generally tolerated as part of the river scene, and sights of their nests on river detritus and of their young are generally enjoyed; at the worst they may cause minor disturbance. Much the same could be said for duck (usually Mallard or Tufted), but swans may pose a more serious problem if weed growth is poor and the density of birds too high. Most keepers would accept that a pair of swans on any single beat is desirable, but sometimes, often due to pressures or disturbance in other areas, there may be as many as fifty to a hundred juvenile birds on a single reach. This causes excessive disturbance and too much weed may be removed. Increasing numbers of Canada Geese are also giving rise to concern and it may be that some effective control will soon have to be used. Canada Geese can legally be shot during the statutory season for wildfowl.

The fish eaters present another serious problem. Few would not welcome the Kingfisher which is still quite common where suitable nesting banks are to be found. The Dabchick may cause a small degree of disturbance, but much more serious are the larger predators, the Heron and Cormorant. Decline in stocks of coastal marine fish and increases in fish farming and in the stocking of lakes may have caused cormorants to change their feeding habits and move progressively inland. Analysis of droppings under trees where the cormorants roost have shown without any doubt that fish, particularly salmon parr, form a part of the bird's diet.

Unfortunately control is a real problem as shooting of swans, herons and cormorants (as well as most other birds) is prohibited by law unless a special licence is obtained. Bird scaring devices (explosive cartridges) are only of limited value and would cause too much disturbance to the river environment.

The third group of birds making use of the riverine habitat includes those that feed on insect life, such as swallows, martins and swifts, and those that feed on invertebrates living in the marginal vegetation. Many other birds feed on fruits and seeds.

Large numbers of birds may also have their nesting sites beside the river and management should leave these areas undisturbed during the nesting season.

Mammals

Fenland is amazingly rich in mammals large and small; where the surface vegetation and the topsoil has been renoved it is riddled with small mammal runs. Of the small species, Wood Mice, Field Voles and Water Voles are herbivorous. The last is often seen in and around the river and has a diet that includes the stems of reeds and rushes. On many rivers it is a declining species, due to reducing habitat and predation by mink. The Itchen has one of the best populations of Water Voles in the United Kingdom.

The Water Shrew, Common Shrew and Pygmy Shrew are all carnivorous, but are seldom noticed. Fishermen are more likely to notice the considerable variety of bats that frequent the river at dusk. These may nest in old trees, barns, riverside buildings and even under bridges; they are protected and it is good to see so many species of bat thriving in the chalk stream area.

The larger mammals may also be grouped by their feeding habits. Leaving aside the herbivorous rabbit, hare and squirrel, the rat is of importance, not least because it is a carrier of Weil's disease. It is important that food is not left on the river bank and in fishing huts. Stoats may be relatively rare on some reaches, but in others may pose a problem through their depredations on game bird eggs and chicks. Mink were thoroughly established on the Test and Itchen but intensive trapping using the Game Conservancy mink rafts has effectively eliminated them from the Itchen. Mink can also cause damage to fish stocks, particularly where the fish are confined in areas like stews.

Otters are staging a recovery nationally and reintroduction has been successfully carried out on a few sites. Old carriers with suitable holt trees and islands should be left undisturbed. Many mortalities occur on roads; underpasses should be made with fencing to force the animal through the river or culvert rather than over the road. The construction of artificial holts in quiet carrier streams can also help. Their territorial range is large; they may travel thirty kilometres in a day and it is vital that suitably quiet areas occur throughout their territory. Their diet includes quite large numbers of eels and few keepers mind losing a few trout for the sight of this fascinating and unhappily now rather rare inhabitant of the chalk stream.

Although otters have been successfully reintroduced, the introduction of new and alien species must always be viewed with great suspicion. Sometimes, as in the case of the European Beaver, the species may have existed in the area hundreds of years ago, but this should not be seen as a reason for its reintroduction. The habitat and land management techniques have changed and it is dangerous to assume that the introduced species will necessarily settle successfully into the eco-system.

The wide ranging territories of many of these species emphasises yet again the importance of the holistic approach of regarding the whole river and its surroundings as one highly complex and variable ecosystem with a great deal of mutual interdependence.

References

Brewis, A., Bowman, P. , Rose, F. *The Flora of Hampshire,* Harley Books with Hampshire and Isle of Wight Wildlife Trust, 1996.

Mühlberg, Helmut, *The Complete Guide to Water Plants*, E.P. Publishing Ltd., 1982.

Spencer-Jones, David, and Wade, Max, *Aquatic Plants*, ICI, 1986.

Stace, C.A., *New Flora of the British Isles*, Cambridge University Press, 1991.

53 Mayfly, *Ephemera danica,* male spinner *Photo: Warren Gilchrist*

Chapter Four

CHALK STREAM INVERTEBRATES

Nearly all the food of game fish (trout, salmon and grayling) in a chalk stream consists of the river invertebrates, which must have the conditions in which they can thrive. The conservation of invertebrates is bound closely to the welfare of the river, and the weed beds which shelter and feed them. The care of these creatures, mostly highly sensitive to pollution, should be a prime concern of all river managers, not only for trout, but for the welfare of all life in the river.

This chapter summarises the most significant river invertebrates, and discusses the conditions that they need.

While scientific keys to identification are beyond the scope of this book, some guidance on simple recognition is included. Those wishing to take identification further can refer to the keys listed in Appendix B.

There are many unknown factors about the requirements, life habits and success of the river invertebrates and some of these problems are listed, to suggest where research might be directed in future. Some relatively simple monitoring can be carried out which will build up a useful database by which fly numbers, for example, can be reliably compared over the years and which will draw attention to any sharp decline in river fly life.

The river invertebrates

The following invertebrates, not all necessarily aquatic, form part of the diet of trout and grayling. In the case of the insects, this often involves early stages in their life histories.

a. Insects:
    - Upwinged flies (Ephemeroptera)
    - Dragonflies (Odonata)
    - Stoneflies (Plecoptera)
    - Water bugs (Hemiptera)
    - Alder flies (Megaloptera)
    - Butterflies and moths (Lepidoptera)
    - Sedge or caddis flies (Trichoptera)
    - True-flies, including midges, gnats and crane-flies (Diptera)
    - Wasps, bees and ants (Hymenoptera)
    - Beetles (Coleoptera).

b. Crustaceans:
    - Freshwater shrimps (Amphipoda)
    - Hog-lice or water wlaters (Isopoda)
    - Crayfish (Astacidae).

c. Molluscs:
   Snails (Gastropoda)
   Mussels (Bivalvia)
   Water spiders and mites (Chelicerata):
   Water mites (Hydracarina)

d. Worms and Leeches:
   Flatworms (Platyhelminthes)
   Roundworms or nematodes (Nematoda)
   Leeches (Hirudinea)

The most significant foods of game fish

While trout and grayling may eat all the above invertebrates, most form only a minor part of their diet. This chapter concentrates on those which are most important to the chalk stream fish.

The relative significance of each type varies within a river, and with the seasons. The following generally form the greatest part of the diet, not necessarily in this order:

a. The molluscs, and particularly the snails (Gastropoda) and mussels (Bivalvia).
b. Freshwater shrimps (Gammaridae).
c. The upwinged flies (Ephemeroptera).
d. The Sedge or caddis flies (Trichoptera).
e. The True-flies (Diptera).

Particular attention has been paid to the upwinged flies (Ephemeroptera) which hold a particular fascination for anglers. When these are on the surface of the water, or immediately under it, the dry fly comes into its own, bringing to anglers the challenge of identification, imitation and presentation. The same applies, perhaps to a lesser extent, when imitating the larval, or nymph, stage of these insects.

Sensitivity to pollution

In 1981 the Department of the Environment and the National Water Council commissioned a Biological Monitoring Working Party (BMWP) to produce a scale by which the varying degrees of tolerance to organic pollution shown by families of invertebrates can be measured. Those which are most sensitive are given the highest score (10). This scoring system is valuable for fishery managers to show which invertebrates are most likely to be affected if pollution occurs. As many of the upwinged flies and sedge flies have the lowest tolerance to pollution, they may provide the first indication that there might be a problem, and their presence is a reliable indicator of relatively pure water. The monitoring system described below takes this into account. See Appendix C for a full list of BMWP scores.

Life habits and requirements of the most significant invertebrates

The upwinged flies (Ephemeroptera)

These are known by the scientist and many laymen as 'mayflies' – somewhat illogically,

Mayflies (Upwinged flies)

Stonefly – Yellow Sally
Isoperla grammatica

Turkey Brown
Paraleptophlebia submarginata

Mayfly
Ephemera danica

Alderfly
Sialidae

Blue -winged Olive
Serratella ignita

Small Dark Olive
Baetis scambus

Damselfly
Agrion splendens

Yellow May dun
Heptagenia sulphurea

Angler's Curse
Caenis luctuosa

Caddis flies (Sedge flies)

Caseless

Sandfly
Rhyacophila dorsalis

Grey Flag
Hydropsyche instabilis

Dark Sedge
Polycentropus flavomaculatus

Cased

Caperer
Halesus radiatus

Grannom
Brachycentrus subnubilus

Welshman's Button
Sericostoma personatum

Cinnamon Sedge
Limnephilus lunatus

Limnephilus binotatus

Brown Silverhorn
Anthripsodes cinereus

Photos: Dr C. Bennett

because insects of this order may be seen on the wing during much of the year. The angler refers to them as the upwinged flies, reserving the name 'mayfly' for the large *Ephemera* species which are normally on the wing in late May or early June. The terms 'dun' and 'spinner' refer exclusively to members of the Ephemeroptera.

There are some 51 species of Ephemeroptera in Britain, but only about 18 are of major importance in the southern chalk streams. These are summarised in Appendix A.

The nymph, or larva (54)

The nymph hatches from the egg and lives for between two months and a year, or even longer. The nymphs live in a variety of habitats in the river – some burrow into fine gravel or silt on the bottom and some live amongst mosses, aquatic plants or stones. The habits of the Ephemeroptera of most significance in southern chalk streams are as follows:

a. Stone clingers.	The nymphs cling to rocks or stones: *Heptagenia* spp. (e.g. Yellow May dun) (54).
b. Agile darters.	The nymphs dart, in short bursts of swimming, from stone to stone or amongst vegetation. Some *Baetis* spp. (Olives) (54); *Centroptilum* sp. (Spurwings); *Cloeon* sp. (Pond Olives); *Procloeon* sp. (Pale Evening).
c. Bottom burrowers.	The nymphs tunnel into the sand or silt: *Ephemera* spp. (Mayfly) (54).
d. Silt crawlers or creepers.	The nymphs move slowly over sand, mud or vegetation: *Caenis* sp. (Anglers' Curse) (54).
e. Moss creeper.	Rather inactive: *Serratella* sp. (Blue-winged Olive) (54).

The nymphs breathe through their gills, which consist of small plates running down either side of their body. Some insects which live in still water, or very slow-flowing rivers, also use these plates to generate a flow of water over the gills. They grow in stages ('instars'), bursting out of their old skins many times as they grow bigger. Virtually all solid, submerged objects, (including aquatic plants) attract a complex plankton-based layer known as the 'periphyton'. This slimy covering is the principal food of many of the river invertebrates, and especially the Ephemeropteran nymphs. Some species (such as some *Baetidae*) have up to five generations a year, some only one, and the mayflies, it is believed, have a generation either every year, or every other year.

Nymphs have a habit, particularly at night, of drifting downstream from their normal stone or plant and regaining a hold further down river. This is known as nymphal drift. The mechanics of this have been thoroughly studied, but the real reasons for this drift, and the advantage to the insects, are still not fully understood. What are the losses from such a process, particularly when there is relatively fast-flowing water, due to heavy rain or reduced weed? One of the effects of this process is to conceal a local, major pollution incident which wipes out the invertebrates downstream of an escape of insecticide, for example. Nymphal drift will quite quickly replace the species (but not the abundance) lost just below the pollution site, and sites further downstream will take longer, and the further downstream, the longer it will take. Even if the numbers are greatly reduced, it is likely that the BMWP scores (which do not at present take abundance into account) will be back to normal by then and there will be no scientific traces of the disaster. This is why regular monitoring of the river, as described below, is so important.

The dun, or sub-imago

The nymph is ready to hatch into the sub-imago stage when it becomes 'ripe'. That is to say, it ceases feeding and the wing cases become dark. The body fills with gas (giving it a silvery look), which helps it to swim, or crawl, up to, or close under, the surface. Its skin splits and the fully-winged fly emerges on to the surface of the water, resting while its wings dry and it gathers strength to fly off to the safety of the vegetation, where it waits to change into its final stage. This is the anglers' dun, and while it is on the water it is vulnerable to hungry fish and birds. When the weather is dull or wet, the wings take longer to dry and the fly takes longer to reach safety. The wind, too, has an immediate effect on these insects which cannot make headway into even a light breeze and generally fly downwind. Some species crawl onto vegetation instead and are less vulnerable. The insects may remain in the dun stage for between 12 and 36 hours, although the *Caenidae* – the Anglers' Curse – change in a matter of minutes. Duns are recognised by their dull appearance and opaque wings, caused by a coating of tiny hairs that prevents them becoming waterlogged. Many of these duns are called 'olives' by the angler.

The spinner, imago or perfect insect

The perfect insect is the final stage in the life cycle, and transformation from the sub-imago, or dun, usually takes place on the river bank. The new insect bursts out of its old sub-imaginal skin and reveals a beautiful shiny creature, with transparent wings and long tails. The fly is now able to mate, a process that is carried out in flight. The males swarm in a sort of dance, which is why anglers call them spinners. This usually takes place over land. Females either fly close to, or into, the swarm of males, the males grab them with their long forelegs and the two leave the swarm to mate. The male can be distinguished easily by the two claspers at the end of its body by which it holds the female while mating and also the turbinate, upward-looking eyes. The male may copulate several times before dying, usually over the land, which is why male spinners are of less interest to the angler than the females. The mated females either fly over the water to lay their eggs, or to a place of shelter. Those which shelter may do so for quite a long time, depending to an extent on the weather. The Baetidae crawl down a stick or weed into the water. The Blue-winged Olive spinners (*Serratella ignita*) fly upstream over the river with their tails and abdomen bent down and forward, cradling a tiny ball of eggs which has been exuded from their oviduct. The egg ball is either dropped into the water or the female brushes her abdomen into the water (they often seek out rough water to help them) to wash off the egg mass. The female spinners then die and often float down the river exhausted, with their wings outspread. These are the spent spinners that are so much loved by the fish. In general, spinners live for less than a day, but, depending on the species and weather conditions, may live for up to a week.

The eggs

As explained above, these may be laid on the surface of the water. They sink to the bottom and, because they are sticky, remain on stones, gravel, aquatic plants and so on.

Some species, such as the Baetidae, crawl down aquatic plants or sticks into the water and lay their eggs on something under the water. The eggs may hatch after a few days, or many months, depending on the species and on factors such as water temperature. The life cycle is complete.

Identification of the Ephemeroptera

The food of trout is obviously of prime importance if they are to thrive in the river, and the manager of a fishery must therefore have a good idea of the types of food available and the conditions which they need if they are to continue to support a healthy population of fish. The start of all this must be the identification of the various forms of food. Really accurate identification of most types of fish food, though desirable, is not essential, but the identification of the Ephemeroptera is more important than most, because:

a. The Ephemeroptera are important sources of food for the chalk stream trout.

b. The Ephemeroptera are probably the insects most sensitive to pollution in a river. In part, this is because the nymphs feed on tiny particles in the river. Particles of pollution stick to those food particles, which then become too large for the nymphs to eat. If they thrive, then it is likely that the conditions in the river are such that the other types of fish food will thrive also. Of the genera of the Ephemeroptera most important to chalk streams, the Yellow May dun (*Heptagenia sulphurea*), the Blue-winged Olive, (*Serratella ignita*), and the Mayfly, (*Ephemera danica*), are amongst the most sensitive to pollution. It will be seen from Appendix C, the BMWP pollution score system, that the Baetidae appear to be relatively insensitive to pollution. In fact, most species are highly sensitive, although the Large Dark Olive (*Baetis rhodani*) is less sensitive. As the BMWP score is based on the whole family, the score for the Baetidae makes them appear less sensitive than many of them actually are.

c. The upwinged flies probably hold the most interest to the angler, who is likely to wish to know more detail about them than about any other order of invertebrates.

The identification process

Accurate, scientifically acceptable identification of many of the species of Ephemeroptera is difficult and will not always be practicable. If one looks at any of the scientific keys listed in Appendix B, one will see that some of the insects in their dun, and even spinner, stages simply cannot be identified beyond their genera. This is especially true of the Baetidae, which are amongst the commonest of all the Ephemeroptera on chalk streams. If one wishes to identify the Ephemeroptera scientifically and accurately, using the proper keys, the best thing to do is to collect nymphs from the river and identify them under a low-power microscope. (A student stereomicroscope magnifying up to 30 times, available from microscope suppliers, meets the requirement very well.) All species may be identified without confusion by this method, although *Baetis scambus* and *B. fuscatus* are very difficult and others need a compound microscope. If one wishes to see what a particular species looks like in its dun and spinner stages, one can breed the nymphs in a small aerated aquarium (to maintain water movement). When the nymph hatches into the dun stage, it leaves its nymphal shuck behind in an almost perfect state, and this can be used to identify the dun which has hatched. If this dun is kept in a moist container for a while, it will hatch into the spinner stage, which will be similarly available for inspection.

While use of the proper keys is essential for scientific accuracy and records, this is often not necessary for the angler or keeper, who may prefer something easier and more rough and ready.

Identification of the nymphs

Generally, the nymphs of the Ephemeroptera identify themselves as Ephemeropteran by having three tails, gills on the abdominal segments and only one claw on the end of each leg. (Stoneflies always have two claws at the end of each leg).

Identification can be achieved by the use of scientific keys listed in Appendix B. A low-power microscope will usually be needed. The shape of the gills is a particularly significant feature. The keys can be used, not only with the nymphs, but with the skins of the nymph from which a sub-imago, or dun, has hatched. It may be the only way of positively identifying a female imago or sub-imago.

A few of the nymphs are self-evident, and may be identified at the riverside in a tray or bucket of water:

a. The mayfly, *Ephemera* sp. (55) Large, pale-ochreous yellow, with eyes visible as little black dots. The gills are placed along the back and are continually in motion in order to keep a flow of water over them. Under a microscope there can sometimes be seen little 'lumps' clinging to the skin. These are protozoa (unicellular animals) of the genus *Vorticella*.

b. The Blue-winged Olive, *Serratella ignita* (59). Its most easily-recognised characteristic is the way it swims. Possibly because it is a clinger by habit, and swims little, its swimming action is awkward and looks rather like a bucking bronco. It looks, too, as though it is trying to swim like other nymphs, but has been damaged in some way. The speckles on the legs are more marked than on Baetidae nymphs.

c. The Yellow May nymph, *Heptagenia sulphurea* (57). Because it clings to stones, it is flattened, with large forelegs well adapted for clinging – almost crab-like.

Identification of the duns, or sub-imagines

One can simplify matters by limiting the range of species to those members of the Ephemeroptera most common on the southern chalk streams, and to those of most importance to the fish. An observer can achieve a broadly accurate identification which will often be adequate. The species with which we should be most concerned on southern chalk streams are:

Family	Genus	Species	Synonym	Common name
Baetidae	*Alainites*	*muticus*	*pumilus*	Iron-blue
	Baetis	*fuscatus*	*bioculatus*	Pale Watery
	Baetis	*rhodani*		Large Dark Olive
	Baetis	*scambus*		Small Dark Olive

	Baetis	*vernus*	*tenax*	Medium Olive
	Centroptilum	*luteolum*		Small Spurwing
	Cloeon	*dipterum*		Pond Olive
	Nigrobaetis	*niger*		Iron-blue
	Procloeon	*bifidum*	*pseudorufulum*	Pale Evening
	Procloeon	*pennulatum*		Large Spurwing
Caenidae	*Caenis*	*macrura*	*halterata*	Anglers' Curse
	Caenis	*luctuosa*	*moesta*	Anglers' Curse
	Caenis	*rivulorum*		Anglers' Curse
	Caenis	*pusilla*		Anglers' Curse
Ephemerellidae	*Serratella*	*ignita*		Blue-winged Olive
Ephemeridae	*Ephemera*	*danica*		Mayfly: Green Drake
Heptageniidae	*Heptagenia*	*sulphurea*		Yellow May
	Rithrogena	*semicolorata*		Olive Upright
Leptophlebiidae	*Paraleptophlebia*	*submarginata*		Turkey Brown

Several of these species are more or less self-evident, and so present no real problem of identification. Those which are not self-evident can be classified well enough to be of some practical use to the riverside observer. It is best if the fly is examined in the hand, and a small net with a relatively long handle to catch the insect serves this purpose well. A small folding net, with a telescopic handle, which can be carried in the pocket, may be used. Aquarium suppliers provide a variety of nets, and even the seaside shrimping net on a bamboo cane can achieve all that is necessary.

Once caught, the fly should be put into a glass test tube for identification, but it tends to be so lively that it can be difficult to see close detail. It is best, therefore, to quieten or even kill the insect. The insect can be quietened without killing it by putting it in a fridge for about twenty minutes, or even in soda water. It may be killed by carrying a small 15 ml. bottle containing a piece of blotting paper or cotton wool soaked in ethyl acetate or strong .880 ammonia. (A better method is to put a little plaster of Paris in the bottom of the bottle, which can soak up the chemical each time it is to be used. The insect can be kept clear of the plaster with a disc of dry blotting paper on top of the plaster).

The Ephemeropteran duns which are self-evident

The following Ephemeropteran duns, or sub-imagines, can be identified by a superficial inspection, sometimes even while they are still on the water:

a. **The Mayfly (Green Drake).** *Ephemera danica* (58). These straw-coloured flies are quite the biggest on the water and cannot be mistaken for anything else, either in the dun or spinner stages.

55 Mayfly, *Ephemera danica*, nymph

56 Blue-winged Olive, *Serratella ignita*

57 Yellow May nymph, *Heptagenia sulphurea*

58 Mayfly, *Ephemera danica*, female dun

Photos: Dr C. Bennett

59 Blue-winged Olive, *Serratella ignita,* female spinner

60 Iron Blue dun, *Alainites muticus,* male dun

61 Turkey Brown, *Paraleptophlebia submarginata* female dun *Photos: Dr C. Bennett*

62 Large Dark Olive, *Baetis rhodani,* female dun

63 Pale Watery, *Baetis fuscatus,* male dun

64 Small Dark Olive, *Baetis scambus,* male dun

Photos: Dr C. Bennett

65 Large Spurwing, *Procloeon pennulatum*, female dun

66 Small Spurwing, *Centroptilum luteolum*, male dun

67 Pond Olive, *Cloeon dipterum*, female dun

Photos: Dr C. Bennett

b. **The Iron-blue dun.** *Alainites muticus, Nigrobaetis niger* (60). These are the only small, blue-black up-winged flies, and are quite unmistakable. A close look shows that their wings, body and legs are really dark olive. If they hatch in large numbers, they look like little spots of ink coming down the river.

c. **The Blue-winged Olive dun**. *Serratella ignita* (59). Its large, grey-blue wings tilt the body over at an angle as it floats down the river, making the flies look like miniature yachts. In the hand, its identification can be confirmed by the presence of three tails, although one is often broken off. There are other flies of similar size with three tails, but all are relatively uncommon on the chalk streams.

d. **The Anglers' Curse or Caenis.** *Caenis luctuosa, C. macrura, C. pusilla, C. rivulorum.* There is normally no need to break them down to individual species. They are recognised by their small size, white wings and three tails. They are normally only around in the early morning and in the late evening, and may often be found in spiders' webs.

e. **The Yellow May dun.** *Heptagenia sulphurea*. The only significantly yellow up-winged fly to be seen on the river, but never, in my experience, in large numbers. It is relatively large, but not as large as the mayflies. It is of importance as a fly which is particularly sensitive to pollution. *Rithrogena semicolorata*, the Olive/Yellow Upright, is becoming more common in the south and may be confused with *H. sulphurea*.

f. **The Turkey Brown.** *Paraleptophebia submarginata* (61). An uncommon insect seen in May and June. Its dark mottled wings and three tails make it easily identifiable.

g. **The Large Dark Olive.** *Baetis rhodani* (62). Because of its large size and its appearance in early spring, the Large Dark Olive is pretty well self-evident, although it is discussed below with the other olives.

The Ephemeropteran duns which are not self-evident

We are now left with the species which are not self-evident. They are all olive in colour and have two tails. They can, however, be identified well enough to be of some practical use to the riverside observer. They are the anglers' 'olives', and, superficially, they look pretty much the same. These include species of *Baetis* (the 'olives'), *Centroptilum* (the Small Spurwing), *Cloeon* (the Pond and Lake Olives) and *Procloeon* (the Pale Evening Dun and Large Spurwing). They may be roughly sorted out, however, by closer inspection, because they do provide some clues as to their identity. They may be small, medium or large (hence the terms 'Small Olive' (64), 'Medium Olive' and the largest one, *Baetis rhodani*, the Large Dark Olive). The latter is often self-evident if seen early in the season when other olives are scarce.

Some of these olives may be a noticeably pale olive, and such may well be the Pale Watery, a Spurwing or the Pale Evening dun, which is particularly pale.

The males of the Pale Watery dun (63) have distinctive lemon-yellow eyes. The females look very like those of the Pale Evening dun, but it has hindwings, which separates it immediately from the Pale Evening dun.

The Small Spurwing (66) is a common, if overlooked, fly of the chalk streams. A hand lens may reveal the very small spurwing. It is often seen at rest with its wings slightly open.

The Large Spurwing (65), although seldom seen on the water, normally sits with its wings slightly open, unlike other olives, and this appears to be a characteristic of both spurwings.

The Pale Evening dun. There are clues to the identity of this fly, but perhaps the easiest is that it doesn't have any hindwings, which separates it from the others, and particularly from the Pale Watery dun.

The Pond Olive (67) is unfortunately increasing on the Test. I say 'unfortunately' because it can cope with oxygen deficiency, relatively high water temperatures and low flow rates (as it has moveable gills), which have been a feature of the weather in recent years, and Pond Olive nymphs are often abundant. They have no hindwings.

There is a rough-and-ready key below, which may be used with simple equipment and should help to sort these out. Please note that this key is not scientifically reliable, but may be useful because the scientific keys will not break down the females (♀) below genus level. Use should be made of scientific keys if more accuracy, particularly with the males (♂), is required.

The species are limited to those most likely to be found on the southern chalk streams.

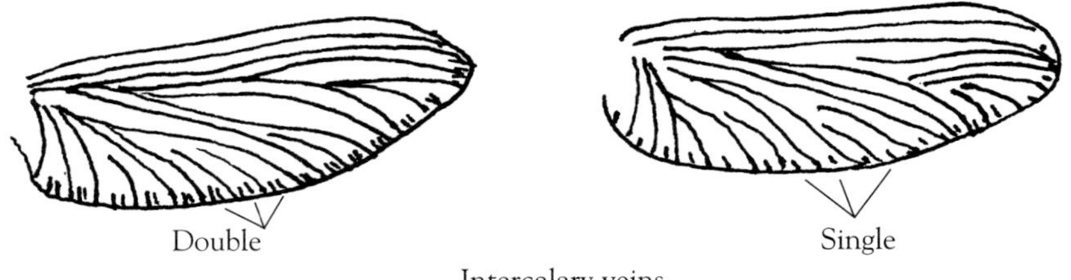

Intercalary veins

Key to Ephemeropteran duns with two tails

1. Intercalary veins
a. Double 2
b. Single 6

2. Sex
a. Male 3
b. Female 5

3. Eyes
a. Red-brown 4
b. Yellow .... **Pale Watery dun ♂**

4. Size
a. Medium large (body 7-10 mm) .... **Large Dark Olive dun ♂**
.... (if no spur on hindwing, *B. atrebinatus*)
b. Medium (body 6-8 mm) .... **Medium Olive dun ♂**
c. Small (body 4-6 mm) .... **Small Dark Olive ♂**

5. Size

a. Medium large (body 7-10 mm) .... **Large Dark Olive ♀**
(Eyes may be very dark green-black) (if no spur on hindwing, *B. atrebinatus)*

b. Medium (body 6-8 mm) .... **Medium Olive dun ♀**

c. Small to medium (body 5-8 mm).. **Pale Watery dun ♀**
(if no hindwing, **Pale Evening dun ♀)**

d. Small (body 4-6 mm) .... **Small Dark Olive ♀**

6. Hindwing

a. Present (but only a small spur).. 7

b. Absent.. 8

7. Size

a. Medium to large (body 8-10 mm) ....**Large Spurwing**
(The wings are normally held open. Rounded tip to hindwing.)

b. Medium to small (body 6-8 mm)..**Small Spurwing**
(Pointed tip to hindwing.)

8. Sex.

a. Male... 9

b. Female... 10

9. Body and eyes

a. Very pale straw, brownish patches on back, last 3 segments pale orange. Eyes dull yellow. **Pale Evening dun ♂**

b. Two red lines through the eyes.. **Pond Olive ♂**

10. Body and eyes

a. Very pale straw, brownish patches on back. Eyes dark olive-green. **Pale Evening dun ♀**

b. Two red lines through the eyes..**Pond Olive ♀**

Identification of the spinners

When the duns hatch into spinners, the colours of both the body and the wings may be very different from those of the dun. Furthermore, the spinner body colours may vary within a species. For example, the Blue-winged Olive spinner ('Sherry spinner') may vary from pale straw-brown to a dark reddish-brown. Many male spinners have pale or white bodies, because their guts are filled only with air, while the eggs of the female spinners give colour to their bodies.

1. Spinners which are self-evident

a. **The Mayfly (Spent Gnat).** *Ephemera danica.* These straw-coloured flies are quite the biggest on the water and cannot be mistaken for anything else, either in the dun or spinner stages.

b. **The Iron-blue spinner. (♀ Little Claret spinner, ♂ Jenny spinner).** *Alainites muticus, Nigrobaetis niger.* These small spinners may be distinguished by the claret (ruby) body of the females and the translucent white body of the male, with its last three segments dark orange-brown.

c. **The Blue-winged Olive spinner (Sherry spinner).** *Serratella ignita* (59). Its identification can be confirmed by the presence of three tails, although one is often broken off (but the broken stump should still be evident). There are other flies of similar size with three tails, but all are relatively uncommon on the chalk streams. The bodies of the female spinners vary from olive-brown to a dark sherry-red, and the eyes are greenish-brown. The males have bright red eyes, with bodies of less pronounced dark sherry-red than the females. The Turkey Brown spinner is very similar, but is much less common.

d. **The Anglers' Curse or Caenis.** *Caenis luctuosa, C. macrura, C. pusilla, C. rivulorum.* There is normally no need to break them down to individual species. They are recognised by their small size, transparent wings and three tails. They are similar to the duns, although their bodies are whiter and their tails are longer.

e. **The Yellow May spinner.** *Heptagenia sulphurea.* The only significantly yellow up-winged fly to be seen on the river, but never, in my experience, in large numbers. It is relatively large, but not as large as the mayflies. It is of importance as a fly which is particularly sensitive to pollution. The wings of the female spinner have pale yellow leading edges. The wings of the male spinner often have smoky-grey leading edges, with a golden-brown body. Both sexes have blue eyes, although the colour tends to change with age (maturity). *Rithrogena semicolorata*, the Yellow Upright spinner (it changes its name from Olive Upright dun when it becomes a spinner!), is becoming more common in the south and may be confused with *H. sulphurea*.

f. **The Turkey Brown.** *Paraleptophebia submarginata.* An uncommon insect seen in May and June. In the spinner stage it is very similar to the Blue-winged Olive.

2. Spinners which are not self-evident.

The rough-and-ready key below, which may be used with simple equipment, should help sort these out. Please note that this is key is not scientifically reliable, but may be useful because the scientific keys will not break down the females (♀) below genus level. Use should be made of scientific keys if more accuracy, particularly with the males (♂), is required.

The species are limited to those most likely to be found on the southern chalk streams.

Key to Ephemeropteran spinners with two tails

1. **Intercalary veins:**
   a. Double 2
   b. Single 9
2. **Sex**
   a. Male 3

b. Female 7

3. Eyes

a. Red-brown 4

b. Yellow .... **Pale Watery spinner ♂**

4. Tails

a. Ringed red-brown .... **Large Dark Olive spinner ♂**

b. Grey or white 5

5. Size

a. Medium (body 6-8 mm.) .... **Medium Olive spinner ♂**

b. Small 6

6. Thorax

a. Shiny black .... **Iron-blue (Jenny) spinner ♂**

b. Very dark brown or black .... **Small Dark Olive (Red) spinner ♂**

7. Tails

a. Ringed red-brown .... **Large Dark Olive spinner (Red) spinner ♀**

b. Grey or white 8

8. Body & size

a. Pale golden-brown to golden-olive, medium to small (body 5-8 mm) **Pale Watery (Golden) spinner ♀**

b. Dark claret-brown, small (body 6-8 mm) ..... **Iron-blue (Little Claret) spinner ♀**

c. Yellow-brown, ageing to deep reddish brown, medium (body 6-8mm) .. **Medium Olive (Red) spinner ♀**

d. Dark brown to deep red-brown, ringed with a paler colour, small (body 4-6mm) **Small Dark Olive (Red) spinner ♀**

9. Sex

a. Male 10

b. Female 13

10. Hindwings

a. Hindwing pointed at tip .... **Small Spurwing spinner ♂**

b. Hindwing rounded at tip .... **Large Spurwing spinner ♂**

c. No hindwing 11

11. Eyes

a. Pale orange with 2 horizontal red lines .... **Pond Olive spinner ♂**

b. Pale yellow to olive-green 12

12. Tails and eyes.

a. Pale olive tails, pale yellow eyes .... **Pale Evening spinner ♂**

b. White, faintly ringed red, eyes olive-green .... **Lake Olive spinner ♂**

13. Hindwings.

a. Hindwing pointed at tip .... **Small Spurwing spinner (Little Amber sp.) ♀**

b. Hindwing rounded .... **Large Spurwing spinner (Large Amber sp.) ♀**

c. No hindwing 14

14. Tails, etc.

a. Olive-grey, body pale golden-olive with reddish blotches .... **Pale Evening (Golden) spinner ♀**

b. Ringed dark brown, eyes dark green with 2 faint horizontal lines.... **Pond Olive (Apricot) spinner ♀**

The sedge or caddis flies (Trichoptera)

The angler's sedge flies, or caddis, the Trichoptera, are an important source of food for fish. It is seldom that the stomach contents of chalk stream game fish do not reveal the distinctive caddis larva cases. They may be seen on the rivers at any time from March to November.

Their life cycle is different from the Ephemeroptera because they have a pupal stage in addition to the egg, larva and imago, but no sub-imago stage. Some females dip their abdomens into the water during flight to drop their eggs, while others crawl down plants or stones to deposit their eggs under water. The eggs are laid in the water, or on the underside of a leaf, in a gelatinous mass. They hatch after 10 or 12 days, and the larvae are of particular interest (54).

Most larvae camouflage themselves by making a transportable case of sticks, stones, sand, bits of leaf and so on, which are bound together by silken threads produced by the larvae, leaving room for them to expose their head and thorax out of one end and their claspers out of the other. The habitat used by the species determines the type of case which they build. For example, the Phryganeidae and Limnephilidae, which make cases of pieces of vegetable matter, choose a habitat where there are rotting leaves on the bottom of the river. (68) Others choose gravel bottoms and cover their cases in fine pieces of sand. Others do not make transportable cases at all (70). Several spin nets to trap their food – pieces of vegetation or microscopic algae carried by the current – while some use their silk threads to make a tangled web to ensnare other animals which they eat. Of these, some spin funnel-shaped webs, supported by stones. The webs act as a trap for minute organisms, which they eat. Sedge larvae are mostly omnivorous, some herbivorous and some carnivorous. William Lunn, the river keeper of the Houghton Club's Test water from 1887 until 1932, recognised that sedge larvae were serious predators of the Ephemeropteran eggs, and devised fly boards, described later in this section, to try to overcome this. The larger carnivorous larvae may eat smaller larvae of other sedge fly species. They remain in this state for, perhaps, seven months to a year, although this varies considerably.

The larva, whether it had a case before or not, spins a case for pupating and seals itself to a stone or vegetation under water. It remains a pupa for periods ranging from days to weeks, during which it does not eat. It emerges, still in its pupal state, by biting its way out of its pupal case and then may swim to the surface using legs as oars, or it may swim to the water edge and haul itself out.

It hatches into its final stage either on a piece of vegetation above the water, or, like the Cinnamon Sedge (*Limnephilus* spp.), common in chalk streams, may hatch on the surface of the water, sometimes with a marked disturbance. They are distinctive insects, superficially not unlike moths, with hairy wings held like a roof over their body. Most caddis species fly in the evening, and may form swarms, but many of the silverhorns with their long antennae form characteristic daytime groups of zigzagging adults close to the water surface. Whereas the Ephemeroptera do not eat in their final stage, the sedge flies can take in liquid, and therefore live longer. The fish normally only take the adult flies when they are hatching or when the females return to the water to lay their eggs.

The identification of sedge flies is not of much practical importance from a fishing point of view – they vary in size and in shades of brown or black, and imitation of species by the angler is unnecessary. The following is a list of the sedge flies most likely to be seen

68 *Micropterna sequax,* Limnephilidae

69 Grannom, *Brachycentrus subnubilus,* larvae

70 *Plectrocnemia conspersa,* Polycentropodidae

71 Grannom, *Brachycentrus subnubilus,* adult

Photos: Dr C. Bennett

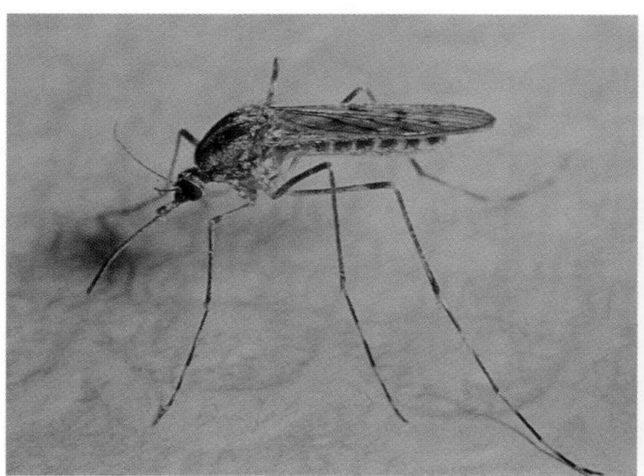

72 Mosquito gnat, *Culiseta annulata*
Photo: G.Ellis/Natural England

73 Chironomid fly, *Cricotopus sylvestris*
Photo: Jon Mold

on the southern chalk streams, with some brief notes which may help in identification:

Hydropsyche pellucidula, Grey Flag.
May – Sept. Large. Flies by day. Uncommon. Marks on wing. Spirals on antennae. Uncased larva.

Rhyacophila dorsalis, Sandfly. Apr to Oct. Medium. Flies afternoon/evening. Common. Wing angles. Uncased larva.

Potamophylax latipennis, Large Cinnamon. April – Nov. Large. Flies by day and evening. Fairly common. Like Caperer, but no wing marks. Cased larva.

Halesus radiatus, Caperer. Aug. – Oct. Large. Flies in evening. Uncommon. Slight wing marks. Cased larva.

Brachycentrus subnubilus, Grannom (71). April – May. Small. Flies by day. Local. Dark. Note wing angle. Cased larva.

Mystacides azurea, Black Silverhorn. May – Sept. Small. Flies afternoon/evening. Uncommon. Very long antennae. Cased larva.

Limnephilus lunatus, Cinnamon Sedge. June – Sept. Medium size. Flies by day. Fairly common. Moon mark on wing. Cased larva.

Sericostoma personatum, Welshman's Button. May – Sept. Medium size. Flies afternoon/evening. Fairly common. Mask on head. Cased larva.

Silo nigricornis, Black Sedge. May to Aug. Small. Flies afternoon/evening. Fairly common. Very dark. Cased larva.

Appendix B lists keys which may be used to identify the nine families most sensitive to pollution, all of which have been recorded from the chalk streams. The Lepidostomatidae (which includes the Silver Sedge) are exceptionally sensitive to pollution. About 40 or 50 species, including the Limnephilidae, may inhabit temporary pools. When the adults hatch in the spring, instead of mating and egg-laying immediately, they become quiescent and settle in places like green ash keys, hollow trees, leaf piles, caves and pine trees. They come down in the autumn to complete their life-cycle. This aestivation (summer quiescance) is triggered by day length and 'switched off' by day length and humidity.

Although widely distributed in Britain, the Grannom (*Brachycentrus subnubilus*) tends to be very local. The NRA survey in September 1989 recorded it from the Test at Greatbridge, north of Romsey, southwards, but the NRA study of macro-invertebrates, July 1993, only recorded the species from Longstock, near Stockbridge and Greatbridge. It has become very rare at Longstock since. On the Itchen, it was not recorded above Eastleigh. Recent reports bear this out. It has been recorded from the River Frome in Dorset. It hatches early in the season in large numbers and game fish feed on them with enthusiasm. It is for this reason of particular interest to chalk stream anglers, and anything which can be done to help its survival and increase will be a benefit to the river. The larva lives in a case which, in the early stages of growth ('instars'), is square (69). It is embedded in vegetation. It secures itself to a weed or twig under water, and its method of feeding is unique amongst the Trichoptera. It filters particles from the water stream with its legs, and then passes them to the mouth. It particularly favours *Ranunculus*, and depends for its well-being on the stability of the plant to which it is anchored. If the weed becomes detached or decomposes, the larva cannot feed and must move to another piece. The possibility that weed-cutting may have caused the marked decline in this species since about 1900 was suspected by William Lunn, the keeper of the Houghton water on the Test, and there seems reason to agree with him. Recent reductions in *Ranunculus*, caused by low flows and swans, have made this problem worse. The adult female carries a mass of green eggs in jelly at the end of her abdomen, and lays them by flying upstream and dipping the abdomen into the water to release a few eggs. This habit has led to the species being known as the 'Greentail'.

The true-flies (Diptera)

So far as chalk stream anglers are concerned, the three main families of the Diptera are the Reed Smuts, the Gnats and the Midges. Two other types of true-fly are also mentioned here.

Reed smuts or black-flies, (*Simulium* spp.), lay their eggs in masses of jelly on protruding vegetation or stones or may lay them under water. The eggs hatch after about a week into worm-like larvae which secure themselves to vegetation or stones under water. They filter food with bristles attached to the head. It is worth noting that, if they are dislodged for any reason, they can make a silken 'lifeline' which they attach to their piece of weed so that they can regain it. They spin a pupal case on aquatic plants and pupate under water, breathing through distinctive filaments. When they emerge, they burst out of their case and float to the surface in a bubble of gas. This keeps them dry and the perfect insect flies off immediately. They are only about 2-6 mm. long. They feed on algae and have a role in its control. The females bite both humans and animals, and there is a well-known biting pest in Dorset, the 'Blandford Fly', *Simulium posticatum*. It is the only one of about 1,000 North European and Asian species which specialises in winterbournes and lays its eggs, not in the river, but in its dry bed or bank about half a metre above the water surface. It is active in May and early June. It can be controlled with a bacteria-produced insecticide, used when *S. posticatum* is the only Simuliid on the river, and it kills nothing else.

Reed smut

Gnats (Culicidae) (72) are larger than reed smuts. There are many different species and detailed identification is quite unnecessary. Gnats can be seen flying in swarms very close to the water. The males of many species are believed to present the female at this stage with a bundle of small midges, seeds and so on wrapped in a silk web. The male takes the opportunity to mate with the female while she is unwrapping the parcel.

Non-biting midges (or plumed gnats or buzzers) (Chironomidae) (73) are present in all water bodies, often in enormous numbers. It is striking how many midges hatch out in a fresh water aquarium stocked purely with mud and stones from a chalk stream. This is a very diverse group, but it is worth mentioning that some of the mud-feeding larvae have a variety of colours, including red, which has become known as the 'bloodworm'. The relatively short-lived pupa swims to the surface to hatch. At this stage it is very vulnerable to game fish, particularly if conditions are such that it cannot get through the surface film. The male adults have plume-shaped antennae. It does not have particular significance for anglers, but is an important part of the chalk stream fish's diet.

Crane-flies, or Daddy-long-legs (*Tipulidae*) may be either aquatic or terrestrial, but they all look roughly similar and get blown on to a river from time to time. The larvae feed in plant roots, wood or leaf litter, or, in the case of the aquatic species, in litter at the bottom of rivers. Some aquatic species are carnivorous.

74 Small crane-fly *Dizzard/Natural England*

75 Crane-fly, *Tipula paludosa*
Photo: Denis G/Buglife

The Hawthorn Fly, (*Bibio marci*), although terrestrial, is one of the earlier flies which anglers see. (It is named after St Mark's day, April 25th). It is instantly recognised by its long trailing legs when in flight. The larvae are often gregarious and are much loved by pheasants. They pupate in cells underground. They are seen round bushes near the riverside and only come on to the river if blown there. They tend to hatch in large numbers, so if the wind is right and the sun is out, some do end up on the river, and fish feed on them with enthusiasm.

Hawthorn Fly, *Bibio marci*

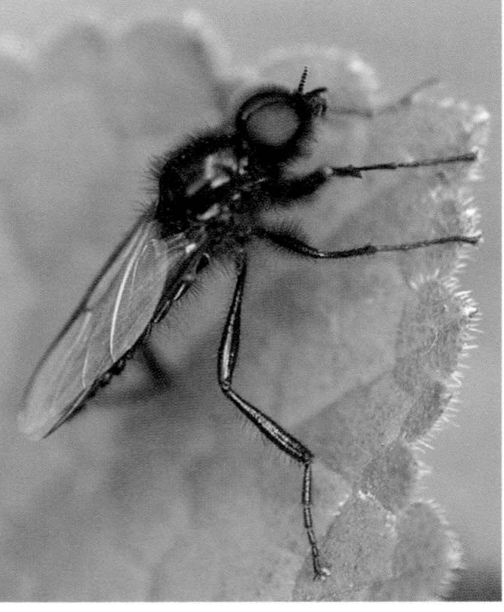

76 Hawthorn Fly, *Bibio marci*
Photo: Northeast Wildlife/Buglife

Crustaceans

There are three types of crustacean of importance to chalk streams:

a. **Freshwater shrimps** (Amphipoda). These include the freshwater shrimp, *Gammarus*, which is abundant and forms a very important part of a game fish's diet. They live on the bottom of the river amongst aquatic plants (starwort and water cress hold them well), stones and mud, always restless and swimming sometimes upside down, sometimes the right way up and sometimes burrowing sideways on the bottom. They are carriers of the larval stage (cystacanths) of *Acanthocephala* parasites, the spiny-headed worms. The orange spot sometimes seen in these shrimps is caused by the parasite *Polymorphus*, whose final host is duck. The adults of the other worms which may be carried by the shrimps (including the pale yellow/orange *Pomphorhyncus*) are parasitic in birds and fishes. The well-being of water plants which these animals like to live amongst is very important to the general welfare of the river.

77 Freshwater shrimp, *Gammarus*
Photo: John Mason

78 Freshwater Louse, *Asellus aquaticus*
Photo: Craig Macadam

b. **The hog-lice or water slaters** (Isopoda). *Asellus*, the freshwater louse, hog-louse or water slater, is not so common or so active as *Gammarus*, but is relatively resistant to pollution. It cannot swim, but crawls amongst the stones and debris at the bottom of the river. It is probably not of much significance as game fish's food, but, if abundant, may be indicative of a polluted river. The spiny-headed worms of the genus *Pomphorhyncus*, and others, parasitize on the hog-lice.

c. **Crayfish** (Astacidae). Although not of significance as food for game fish (although larger fish will eat them avidly), these lobster-like creatures are of importance in clearing up decaying matter from the river bed, thus reducing the river's need for oxygen. They do also compete to some extent with game fish for food. They are chiefly nocturnal, hiding in holes or under stones by day. Stocks of our native species, the White-clawed Crayfish, *Austropotamobius pallipes* (79), have been seriously depleted since 1981, and many rivers in the south of England have lost them completely.

In the mid-1970s the North American Signal Crayfish, *Pacifastacus leniusculus* (80), was introduced to this country from Sweden. (Two other species have also been introduced, but are limited chiefly to ponds and lakes). This species has reached plague proportions on some rivers. It is aggressive, it damages the banks and also eats some of the caddis larvae. It is not thought that this accounts for the current scarcity of caddis flies, because the scarcity exists on rivers which do not have the signal crayfish, but it must have an effect in signal crayfish rivers, bearing in mind the large number of these animals in parts of some southern rivers.

79 White-clawed Crayfish, *Austropotamobius pallipes* *Photo: Environment Agency*

80 Signal Crayfish, *Pacifastacus leniusculus* *Photo: Pete Sibley*

The Signal Crayfish carries a fungus, *Aphanomyces astaci*, which causes crayfish plague, although it is largely immune to the plague itself. The native White-clawed Crayfish, however, is highly susceptible to this virulent fungus disease. As a result, it is now on Schedule 5 of the Wildlife and Countryside Act, 1981, giving it protection against 'taking and sale' in Britain.

If river managers find crayfish, they will be interested to know whether they are the native species, or the aggressive Signal Crayfish. The most significant differences for identification purposes may be summarised as follows:

Rostrum – The sides converge towards the base of the small triangular apex. Signal (N. American Crayfish): sides more or less parallel, sloping down to prominent shoulders some way from the tips. Apex very pointed and prominent.
Body – White-clawed (native) Crayfish: prominent spines on shoulders of carapace, just below cervical groove. (Present in juveniles as small projections). Signal (N. American crayfish): no spines on shoulders of carapace.
Claws – White-clawed (native) Crayfish: top side rough. Underside dirty white, although in juveniles may be pink. Signal (N. American Crayfish): large. Smooth all over; red underneath. On top side, white to turquoise patch at joint of moveable and fixed finger, which gives the Signal Crayfish its common name.

Snails, Limpets and Mussels (Gastropoda and Bivalvia)
Freshwater snails (Gastropoda) are abundant in chalk streams and form an important part of a fish's diet. The pink flesh of wild game fish is obtained from a pigment in these molluscs, as well as from freshwater shrimps, *Gammarus*. They feed mostly on plants or on algae on the surface of underwater objects, using a rasping tongue. They lay eggs in jelly on plants under water. Identification, although quite easy for full-grown adults, is quite unnecessary for river management. The river limpet (Ancylidae) is eaten by game fish and the distinctive shells are sometimes found in their stomachs. The NRA survey of macro-invertebrates on the rivers Test and Itchen recorded the Neritidae, Valvatidae, Hydrobiidae, Lymnaeidae, Physidae, Planorbidae and Ancylidae families of aquatic gastropods. Amongst these are the Ramshorn Snails, (82, 83) the Pond Snails, the Bladder Snails and Limpets. The Succineidae and Zonitidae families from terrestrial, but marshy, habitats, are also recorded.

81 Pond Snail, *Lymnaea peregra*
Photo: David Holyoak

82 Great Ramshorn Snail, *Planorbarius corneus*
Photo: Pryce Buckle

83 Shining Ramshorn Snail, *Segmentina nitida*
Photo: Natural England

84 Pea mussel, *Pisidium* sp.
Photo: Natural England

85 Southern Damselfly, *Coenagrion mercuriale* *Photo: Dennis Bright*

86 Banded Demoiselle, *Calopteryx splendens*, nymph *Photo: Dr C Bennett*

87 Banded Demoiselle, *Calopteryx splendens*, adult
Photo: Dennis Bright

Mussels and cockles (Bivalvia). These are seldom seen because they burrow into the mud, although the Sphaeriidae (84), (freshwater cockles and pea mussels), will sometimes feed on or amongst plants. Amongst the bivalves, the River Test supports ten species of pea mussel (*Pisidium* species). Many of these are eaten by trout whose stomachs can sometimes contain large numbers of these small molluscs.
One of these, the fine-lined pea mussel *Pisidium tenuilineatum*, is a very tiny species with a shell rarely reaching 2mm across and bearing regularly spaced fine lines. This is a mussel that lives very locally in clean hard-water rivers. It typically lives in areas of rivers where the water flow is reduced to allow the build up of fine muddy sediments in which it burrows. *P. tenuilineatum* is a very locally distributed species in the River Test with a population recently discovered living in the Leckford reach. This species appears to be adversely affected by poor water quality arising from nitrate and phosphate enrichment (originating from treated sewage effluent, agricultural run-off, fish farm discharge and intensively managed water cress beds).

Other invertebrates
Although the orders of chief interest to river managers have been described in some detail above, it is worth briefly summarising the following:

a. **The dragonflies and damsel flies** (Odonata). The beautiful adults form no part of the game fish's diet, but the nymphs of the damsel flies do, especially when they come to the surface to hatch. The larvae of the dragonflies are fiercely carnivorous, using a pair of pincers in front of their heads to capture even young fishes and tadpoles. These insects are characteristic of the river's margins and back waters. One can tell the difference between dragonflies and damselflies by the wings and the body:
   i. The wings of dragonflies are held horizontally at rest. Damselflies hold their wings over their backs at rest.
   ii. Dragonflies can hover: damselflies cannot.
   iii. The bodies of dragonflies are large and quite stout: those of damselflies are slender.

   Two species of damselfly are worth mentioning. Firstly, the Southern Damselfly, *Coenagrion mercuriale* (85), which has been reported from both the Test and Itchen valleys. It is listed in the Red Data Book of rare species. Secondly, the Banded Demoiselle, *Calopteryx splendens* (86, 87), which has remained abundant on the southern chalk streams while nearly all the other insects have become much reduced.

b. **The stoneflies, or hard-winged flies** (Plecoptera) (93). These are of little importance to chalk stream anglers, but the Yellow Sally (91), Willow Fly, Early Brown and Needle Fly are the most likely to be seen on the chalk streams. The NRA samples of the rivers Test and Itchen record Nemouridae, (which includes the Early Brown), Leuctridae, (which includes the Needle and Willow flies) and Perlodidae, (which includes the Yellow Sally) (92).
   Their life cycle – eggs laid in the water, an underwater nymph and hatching on land – is not dissimilar to the sedge flies except that there is no pupal stage.

88 Greater Water Boatman, *Notonecta glauca*
Photo: Natural England

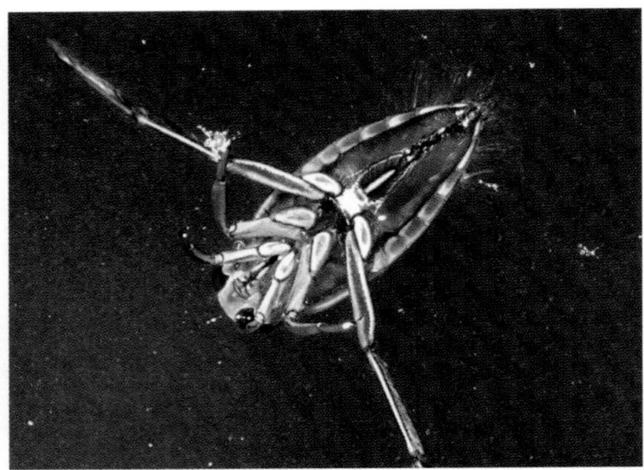

89 Greater Water Boatman, *Notonecta glauca*
Photo: Ben Harmers/Buglife

90 Alderfly, *Sialidae* *Photo: Craig Macadam*

91 Stonefly, *Isoperla grammatica,* nymph
Photo:Dr C Bennett

92 Stonefly, *Perlodidae*, nymph
Photo: Craig Macadam

93 'February Red' stonefly, *Taeniopterix nebulosa,* adult *Photo: Dr C Bennett*

Their presence is another indicator of a relatively unpolluted river, and the Perlodidae are exceptionally sensitive to pollution.

c. **The water bugs** (Hemiptera). These include the water crickets (Veliidae), water boatmen (Corixidae and Notonectidae) (88, 89), the saucer bugs (Aphelocheiridae) and pond skaters (Gerridae), all of which have been reported from chalk streams. The water crickets and pond skaters live on the surface of the water, preying on insects trapped in the surface film. Others, including the common water boatman, live in the water. They are mostly carnivorous and may eat water insects, fish eggs, mussels and even tadpoles and small fish. Some can give a human a nasty 'bite' with their piercing rostrum. The saucer bugs lie buried in gravel or clinging to stones in fast-flowing water. They do not need to come to the surface for air, because they have a type of gill, or plastron, on the underside of their bodies. Game fish will feed on water bugs, and a report by Pentelow on the Food of The Trout in the *Journal of Animal Ecology*, November 1932 found that over 27% of a sample of Itchen trout had Corixidae species of Lesser Water Boatmen in their stomachs.

d. **The alderflies** (Megaloptera) (90). Only the carnivorous larva (54) is aquatic, where it lives in silt or under stones. It is eaten by game fish. The adult seldom falls on to the river.

e. **The butterflies and moths** (Lepidoptera). Nearly all are terrestrial in all stages, although the China Mark moths (*Nymphula* spp.) have some aquatic stages: their larvae feed on some water plants, particularly water lilies. The larvae of the Water Veneer moth, *Acentria ephemerella,* live under water and the adult females normally never rise above the surface of the water. The males can sometimes be quite common in the summer on chalk streams.

There are no butterfly habitats which are unique to chalk streams, but typical riverside vegetation will provide the needs of many of them. Commonly seen by the river are the Peacock, *Inachis io*, the Small Tortoiseshell, *Aglais urticae*, the Red Admiral, *Vanessa atalanta*, the Orange Tip, *Anthocaris cardamines*, the Brimstone, *Gonepteryx rhamni*, Meadow Brown, *Maniola jurtina*, Gatekeeper, *Pyronia tithonus* Ringlet, *Aphantopus hyperantus*, the Comma, *Polygonia c-album*, some of the skippers and the various whites. One may be lucky and see the Painted Lady, *Vanessa cardui*, various blues and even a Clouded Yellow, *Colias croceus*.

Of the moths, the Scarlet Tiger moth, *Callimorpha dominula*, is the most striking. It flies by day and may often be seen in mid to late summer by chalk streams. One's initial attention is usually drawn to it by a flash of bright scarlet. One of its main food plants (but by no means the only one) is Comfrey, *Symphytum officinale*, which may be seen growing close to the river bank, and its black and yellow larvae may be seen feeding openly on the leaves in spring. Another day-flying moth, the Silver Y, *Plusia gamma*, can be seen as soon as it arrives in the summer from the continent, moving rapidly between low-growing plants. Occasionally, numbers may be very high. Unlike the butterflies, there are several moths which are dependent on riverside plants which favour wet or damp soil. Butterbur, *Petasites hybridus*, Hemp Agrimony, *Eupatorium cannabinum*, Common Reed, *Phragmites australis* (food plant

of the Reed Dagger, *Simyra albovenosa*, a nationally scarce moth), Great Reedmace, *Typha latifolia*, ('bulrush'), Hemlock Water-dropwort, *Oenanthe crocata*, and willows are examples of plants which support a variety of moths.
The way a chalk stream is managed can be very significant for Lepidoptera, which need a wide range of plants as food for the larvae.

f. **The wasps, bees and ants** (Hymenoptera). None, except for a very few microscopic parasites, are aquatic, and bees and wasps only occasionally end up on the river. Wasps can, however, choose the river bank for a nest site, which may have to be controlled. A proprietary product may be used in accordance with the maker's recommendations, taking special care not to cause any pollution of the river during the operation. The Environment Agency should be consulted. The flying stages of ants can be blown on to the river in large numbers when they hatch, and game fish feed on them. There is one common parasite, the ichneumon *Agriotypus armatus*, which parasitizes members of the sub-family Goerinae. It overwinters in the caddis fly's pupal case under water, using a small tube protruding out of the case as a plastron, or type of gill. It emerges from the pupal case in the spring.

g. **The beetles** (Coleoptera). Beetles are quite often found in trout's stomachs, but most of them are terrestrial and only reach the water if they have been blown on or have fallen from an overhanging bush. Some (less than 10%) are aquatic in all or part of their life history. Most adults which live under water keep a store of air under their wing-cases from which to breathe, coming to the surface from time to time to replenish the store. The diving beetles (Dytiscidae) and riffle beetles (Elminthidae) are quite common in chalk streams. One striking beetle which may been seen on the riverside is a bright green, iridescent beetle, *Chrysolina menthrastri*, which feeds on water mint.

h. **Worms**. There are many species of aquatic worm, of which the following have been recorded in chalk streams in the south of England:
Flatworms (Platyhelminthes). The Planariidae and Dendrocoelidae, which are carnivorous worms, live on the undersides of floating leaves or beneath stones.
Roundworms or nematodes (Nematoda). These are very small worms, mostly microscopic, which live in sediment on the bottom of a river.
Segmented worms (Oligochaeta). These are mud-dwellers and have a variety of sizes. They are mostly pink or reddish-brown.

i. **Leeches** (Hirudinea). Piscicolidae, the fish leeches, Glossiphoniidae and Erpobellidae. The last two families are parasitic on a variety of other creatures, including molluscs, mammals, amphibians, water fowl, worms and slugs.

j. **The water spiders and mites** (Chelicerata). The water spiders are seldom seen on rivers, but the water mites (Hydracarina) are present on chalk streams and fish do feed on them. They have little significance, however, for river management.

Times of emergence

A rough guide to the times when the Ephemeroptera are likely to be seen on the river is given in Appendix E. There will often be exceptions to this, depending upon the weather and the fact that the insects do not read or necessarily obey the text books, but it is hoped that it will provide some useful guidance, which individual observers can modify with experience.

Conservation of Invertebrates in a river

Aquatic plants

Weed beds play such a fundamental part in stream ecology that river managers should give a high priority to aquatic plants – their health and their control. The requirements of the water plants are dealt with elsewhere in this book, and so is weed cutting, but it is appropriate to make one or two remarks here from the point of view of the conservation of the invertebrates. Anyone about to cut weed must be constantly aware of the dramatic upheaval which will take place for those species dependent on plants. Nymphal drift is referred to above, but weed cutting imposes an involuntary drift upon these creatures, who will need refuge downstream where they can re-establish themselves. This is particularly important for the Baetidae who rely for their oxygen on adequate water velocity over their gills. (Some species such as *Cloeon*, are less reliant, as they have moveable gills). When they float downstream with the current, that flow of water is denied to them. The particular vulnerability of the Grannom sedge fly to weed cutting has already been referred to, and there are other species equally at risk. Weed cutting also affects the height and flow of water in the river, and this has a direct effect on some larvae and even eggs.

Regular cutting of aquatic plants may be essential for the general health of the river, but the impact it may have on some river invertebrates is not fully understood and must always be borne in mind. Weed cutting should be kept to a minimum. Leaving uncut bars of weed at intervals across the river, where possible, may be one solution, and it may also be possible to make a pre-emptive cut in the autumn in order to prevent vigorous growth later. The river invertebrates are abundant and resilient, but the disaster which many of them face when weed is cut is very significant.

The aquatic plants which are most valuable to the Ephemeropterans are the Water-crowfoot, *Ranunculus* spp., Water-celery, *Berula erecta*, Water-milfoil, *Myriophyllum* spp. and Mare's-tail, *Hippurus vulgaris*.

The nature of the river bed

The invertebrates have found a variety of niches in which to live, and so a variety of habitats within a river is necessary if it is to be host to a wide range of species. This habitat includes a bed with stones and gravel, (some fairly loose, or, at least, easy to climb under), to hold enough silt and detritus on which many of the animals feed. If there is too much silt, however, fish may not be able to spawn, and keepers sometimes rake the gravel to prepare for the spawning season. Research by MAFF, following gravel cleaning on the Itchen in 1992, found that spawning success had improved by as much as ten times. A report by the NRA on the gravel cleaning programme on the Test and Itchen,

July 1993, concluded that none of the cleaning methods with which they experimented had any bad effect on the invertebrate fauna. Water jetting is the preferred method. The bottom must also be capable of holding and feeding the river plants. Emergent and submerged vegetation, gravel, stones and silt all have their part to play. The health of invertebrate life and plants must be regularly monitored (this is referred to later) in case the river bed is no longer providing the necessary habitat.

Fly boards and flints

It was William Lunn who devised the idea of fly boards, in order to protect Ephemeropteran eggs from predators in general and sedge larvae in particular. Noting that the predatory sedge larvae cannot swim, he placed floating boards in the river, tethered to bridges and posts, on which the spinners could lay their eggs. Although this was successful, and was widely copied, the practice seems to have become less common in recent years. There may well be a case for greater use of these boards. They should increase Ephemeropteran fly life in general and because they are easily moved, can be transported to parts of the river where fly life appears to be less abundant. A note of caution, however, needs to be struck here. Although the research is not yet complete, nor its significance for the future fully assessed, a parasite, *Spiriopsis adipophila*, has been found in the Thames Catchment Area by Dr Cyril Bennett, in the bodies of the Mayfly (*Ephemera danica*) and, to a much lesser extent, in the Blue-winged Olive (*Serratella ignita*). There is a danger that parasites and possibly diseases may be spread if fly boards are swapped between rivers and even different stretches of a river. With care, however, more use might be made of fly boards. Lunn used two rough deal boards, one inch thick, 7½ inches wide and 8½ feet long, battening them together to prevent warping. He secured a piece of strong wire to one end and fastened it to a bridge or a stake projecting from the bank. Weed must not be allowed to prevent the water flowing freely underneath, either from cut weed floating down or from weed growing too tall. The boards must be checked regularly and can be moved when there is a good coating of eggs on the underside.

94 Fly boards provide a place for invertebrates to lay their eggs *Photo: Guy Robinson*

Many invertebrates use aquatic plants, sticks or stones on which to lay their eggs, to feed and to live amongst. Several species use objects which protrude from the water to climb up or down when hatching or egg-laying. A river can be improved by ensuring that there is an adequate supply of these. The famous river-keeper, Frank Sawyer, used paving slabs placed vertically in the river to help this process. They acted as another sort of fly board, and are still in use today. Suitable flints may then be deposited into a region of quicker water, and may also be placed in shallow water, sticking out at least three inches above the surface. Floating weed may catch on to these projecting stones, and care must be taken to reduce this to a minimum. They may be best placed close to the banks, where they are less obtrusive.

Sites of Special Scientific Interest (SSSIs)

A number of chalk streams, including the Test, Itchen, Frome, Piddle and Avon, have been designated Sites of Special Scientific Interest by English Nature. These were designated as part of a national series of river SSSIs and are typical of the 'classic' chalk rivers of southern England. A list of 'Operations Likely to cause Damage' (OLDs) has been drawn up, and no such operation may be carried out without first obtaining the permission of Natural England.

Habitat

Careful planning of the habitat through which a river runs can do much to conserve the fly life in the river. The Ephemeroptera, for example, are vulnerable when they hatch into the sub-imago, or dun, and neither they nor the perfect insects are very strong fliers. Their life as adults is short, and their breeding can be seriously affected if they are faced, on hatching, with poor weather conditions. Anything which can be done to help them will be an advantage to the fishery. The most practical help which can be given is to grow trees and bushes. Stands and hedgerows of willow, alder and hawthorn some 20 to 25 yards from each bank will provide ideal refuges for the flying insects. While it may be desirable to have a few trees close to the river bank, too many can be a nuisance, and cut out too much light. If the trees are too far away the insects will have an unduly long flight to and from the river and will be vulnerable to birds during their passage. Trees and bushes will harbour terrestrial insects, which will be blown from time to time on to the river. The hawthorn fly, *Bibio marci*, is an example of such an insect, which provides a valuable supplement to the game fish's normal diet. Trees and bushes also act as windbreaks and add significantly, of course, to the beauty of the surroundings. Such trees can be grown cheaply using cuttings placed into the ground in the autumn and protected from predators. The Phytophthora disease of the alder, *Alnus glutinosa*, is increasing, particularly in southern Britain. The Forestry Commission in December 2004 said that "The planting of alder on sites liable to flooding by rivers, on the banks of rivers where diseased alders are known to occur, presents a high risk. While alder is often the most suitable genus for a variety of reasons, owners should take account of the threat of disease and consider other flood-tolerant species, such as willow, as replacements or in mixture." Advice should be obtained from the Environment Agency before carrying out such planting within 8 metres of the river, and of Natural England, too, if the land is part of an SSSI. River banks should not be mown or cleared right up to the water's edge. River margins, back waters and marginal vegetation provide shelter for hatching upwinged flies and an additional habitat for other freshwater insects.

The above considerations support the concept of a protected corridor along the length of the river. Its width, which will depend upon the local topography, may typically be 25 to 50 yards either side of the river bank, but would always include the water itself, the marginal zone between the water and the bank, the bank and part of the adjacent land zone or flood plain. This corridor must be properly managed in order to protect and enhance the conditions which fish, insects, aquatic plants, plants and other wild life need if they are to thrive.

Electro-fishing and fly life

While electro-fishing is a necessary part of predator control in a river, managers must be sensitive to its possible effects on the invertebrates. The nature of the electric field used for electro-fishing is such that larger fish are much more affected than smaller ones. (This can be seen in practice). It follows that the relatively tiny invertebrates are likely to receive a very small current indeed. There were several experiments about 35 years ago in England, New Zealand and America to assess the effect on invertebrates. The experiment in a Lake District river showed that invertebrate drift, particularly of upwinged flies, stone flies and freshwater shrimps, increased by as much as 100 times. The oveall effect was that only about 5% of the population of the stream bed was affected and the scientists concluded that electro-fishing had negligible effect on the river invertebrates. The New Zealand experiment and the American tests also reached the same conclusion, but the Americans sounded an important warning note. Invertebrate drift, although its function is still not fully understood, is almost certainly a crucial factor in the natural mechanisms which regulate a river's invertebrate population. Repeated or regular electro-fishing would interfere with this drift, and it would be wise to confine operations to once-, or at most twice-yearly sessions.

Channel and bank works

While extensive work on the river channel and the banks will be necessary from time to time, the effects on the invertebrates must be carefully considered beforehand. Such operations should be used to protect, rehabilitate and create wildlife habitats, to copy and simulate the best bits in a river and to work with indigenous plants and materials. The approval of both the Environment Agency and, if the river is an SSSI, Natural England, must be obtained before such operations are carried out.

Pollution

The responsibility for pollution control lies with the Environment Agency, which has the necessary expertise and equipment. River managers, however, have a vital role to play, and can do so without needing to use advanced chemical analyses and so on. This is because many of the invertebrates are highly sensitive to pollution, and reduction in their numbers can provide the first indication that all is not well with the river. In general, the upwinged flies, the stoneflies and sedge flies are the most sensitive to pollution, while snails, worms and midges are most tolerant. This means that any noticeable reduction in river fly life may be an invaluable indicator of pollution. A simple method of monitoring against a sudden or significant reduction in invertebrate life has been devised in conjunction with the Environment Agency, and is becoming standard practice in an increasing number of rivers throughout the country.

The abundance of fly life

Reports on the abundance of fly life, even in different parts of the same chalk stream, vary considerably, but there is good evidence that fly life is much scarcer than it used to be. It is not known if the reductions are merely seasonal or cyclical problems, or whether it is a long-term trend. The problem is very complicated, and some of the factors which make it so difficult are listed below:

a. The general lack of a data base by which to compare the present with the past. The NRA's work in September 1989 and in 1991-1992 provide a useful data base for comparison, but it is necessarily limited in the number of samples taken and the methods used for sampling.

b. There are quite strong annual cycles, even of invertebrates such as underground shrimps, which one might think of as not being prone to many changes.

c. The fluctuations in water level both during a season, and from year to year. Exactly what effect does this have, and how long does it last? To what extent is abstraction responsible for this? Exactly what effect does a reduction or increase in water flow have? It is accepted that *Ranunculus* thrives in fast-flowing water.

d. Variations in the time when food is available. The larvae may hatch, one year, at a time when the food development has been delayed, or vice versa, and many larvae may die.

e. Variations in the type of food available. On chalk streams, the Blue-winged Olive, for example, normally has two generations in a season, and its success will depend partly on the type of food available. It is known that the Blue-winged Olive nymph does better when feeding on diatoms than it does on detritus.

f. Variations in abundance of certain food within a river. For example, algae do less well in low light. Midge larvae are unlikely to be so prolific under a shady tree.

g. Are the aquatic plants doing well? If not, which species are suffering? What changes have there been in the water weed species? *Ranunculus* likes fast-flowing water and dies back in low water conditions. Predation by swans of an already reduced *Ranunculus* population is a cause for concern. This is a particular favourite of the Baetidae, whose numbers are likely to suffer without it.

h. Variations in water temperature in different parts of the same river. It tends to be colder nearer to the source of the river.

i. What is the effect of weather? For example, in May 1972, gales in Ireland for nine successive days drowned millions of emerging Ephemera, which did not re-appear for many years.

j. Farming practices. The provision of water meadows as pasture (as opposed to arable) and the provision of buffer strips is beneficial to fly life, provided that cattle are denied access to the river.

k. State of the water – pollution from phosphates, nitrates and so on. Pollution from oil makes it impossible for hatching nymphs to break through the surface film, while pollution from detergents reduces the surface tension to the extent that the hatching nymph cannot obtain enough adhesion or 'grip' to be able to hatch.

l. Depth of the water. For example, most sedge larvae only thrive in water less than 10 feet deep.

m. Behaviour of the insects themselves. Their swarming behaviour, and its dependence on weather (especially wind), landmarks and so on. The effect on competition between species. Their susceptibility to predators and parasites.

Records and recording

If regular records are to be kept at all by river managers, they must be as simple as possible. The range of information which might be logged is daunting, and almost unending, such as water temperature at different levels, air temperature, barometric pressure, water flow at different levels, some thousands of different species, aquatic plants, light intensity and so on, without even attempting to solve all the factors listed above. Because of the potential complications, there is a danger that no records will be kept on the grounds that it is all too difficult.

There is value, however, in a much simpler approach which could, over the years, provide a data base from which preliminary conclusions can be drawn. For example, weed-cutting records on the Allen river in Dorset showed that, whereas it had been necessary many years earlier to cut the weed, say, three times a year, it was now only necessary to do so, say, once a year (Dr M Ladle, *pers. comm.*). This relatively simple record provided a clear indication that there was something wrong with weed growth, and scientists could then move in to investigate. Another example is the continuing work of Alan Frake and Peter Hayes to record the trend of river fly life abundance which relies chiefly on reports from riparian owners, keepers and anglers rather than scientific measurements. In the same vein, a simple monitoring system has been devised in conjunction with the Environment Agency, and is becoming standard practice in an increasing number of rivers throughout the country.

In essence, monthly samples are taken of just four types of Ephemeropteran nymph – the Mayfly, *Ephemera danica*, the Blue-winged Olive, *Serratella ignita*, the Yellow May dun, *Heptagenia sulphurea,* and *Baetis* nymphs generally. Caddis larvae – cased and uncased – are also recorded, as well as the presence of freshwater shrimp (*Gammarus*) and stonefly nymphs. The level of identification is very simple and can easily be learned on one of the monitoring courses run by the Riverfly Partnership and the John Spedan Lewis Trust for the Advancement of the Natural Sciences. In addition to providing early warning of real trouble (when the Environment Agency will come in to help), the records should build up a data base which will allow conclusions to be drawn as to how fly life compares with previous years. A description of this method is included at Appendix D.

Appendix A
Checklist of the British Ephemeroptera

Note. Those of most significance to southern chalk streams are in bold

Family	Genus	Species	Anglers' name
AMELETIDAE	AMELETUS Bengtsson, 1865	*inopinatus* Eaton, 1887	-
ARTHROPLEIDAE	ARTHROPLEA Bengtsson, 1909	*congener* Bengtsson, 1909	-
BAETIDAE	**ALAINITES Waltz & McCafferty, 1984**	***muticus* (Linnaeus, 1758)**	**Iron Blue**
	BAETIS Leach, 1815	*buceratus* Eaton, 1870	Iron Blue
		***fuscatus* Linnaeus, 1761**	**Pale Watery**
		***rhodani* Pictet, 1845**	**Large Dark Olive**
		***scambus* Eaton, 1870**	**Small Dark Olive**
		***vernus* Curtis, 1834**	**Medium Olive**
CENTROPTILUM	**Eaton, 1869**	***luteolum* Müller, 1776**	**Small Spurwing**
	CLOEON Leach, 1815	***dipterum* Linnaeus, 1761**	**Pond Olive**
		simile Eaton, 1870	Lake Olive
	LABIOBAETIS Novikova & Kluge, 1987	*atrebinatus* (Eaton, 1870)	Olive
	NIGROBAETIS Novikova & Kluge, 1987	*digitatus* (Bengtsson, 1912)	Iron-blue
		***niger* (Linnaeus, 1761)**	**Iron-blue**
	PROCLOEON Bengtsson, 1915	***bifidum* (Bengtsson, 1912)**	**Pale Evening**
		***pennulatum* (Eaton, 1870)**	**Large Spurwing**
CAENIDAE	**BRACHYCERCUS Curtis, 1834**	*harrisellus* Curtis, 1834	Anglers' Curse
	CAENIS Stephens, 1835	*beskidensis* Sowa, 1973	Anglers' Curse
		horaria Linnaeus, 1758	Anglers' Curse
		***luctuosa* (Bürmeister, 1839)**	**Anglers' Curse**
		***macrura* Stephens, 1835**	**Anglers' Curse**
		pseudorivulorum Keffermüller, 1960	Anglers' Curse
		***pusilla* Navüs, 1913**	**Anglers' Curse**
		***rivulorum* Eaton, 1884**	**Anglers' Curse**
		***robusta* Eaton, 1884**	**Anglers' Curse**
EPHEMERELLIDAE	**EPHEMERELLA Walsh, 1862**	***notata* Eaton, 1887**	**Yellow Evening**
	SERRATELLA Edmunds, 1959	***ignita* (Poda, 1761)**	**Blue-winged Olive**
EPHEMERIDAE	**EPHEMERA Linnaeus, 1758**	***danica* Müller, 1764**	**Grey/Green Drake**
		lineata Eaton, 1870	Mayfly
		vulgata Linnaeus, 1758	Mayfly
HEPTAGENIIDAE	ECDYONURUS Eaton, 1868	*dispar* (Curtis, 1834)	Autumn
		insignis (Eaton, 1870)	Large Green
		torrentis Kimmins, 1942	Large Brook
		venosus (Fabricius,1775)	False March Brown
	ELECTROGENA Zurwerra & Tomka, 1985	*affinis* (Eaton, 1883)	-
		lateralis (Curtis, 1834)	Dark Yellowstreak
	HEPTAGENIA Walsh, 1862	*longicauda* (Stephens, 1836)	-
		***sulphurea* (Müller, 1776)**	**Yellow May**
	KAGERONIA Matsumura, 1931	*fuscogrisea* (Retzius, 1783)	Brown May
	RHITHROGENA Eaton, 1881	*germanica* Eaton, 1885	March Brown
		***semicolorata* (Curtis, 1834)**	**Yellow Upright**
LEPTOPHLEBIIDAE	HABROPHLEBIA Eaton, 1881	*fusca* (Curtis, 1834)	Ditch
	LEPTOPHLEBIA Westwood, 1840	*marginata* (Linnaeus, 1758)	Sepia
		vespertina (Linnaeus, 1758)	Claret
	PARALEPTOPHLEBIA Lestage, 1917	*cincta* (Retzius, 1783)	Purple
		***submarginata* (Stephens, 1835)**	**Turkey Brown**
		werneri Ulmer, 1919	-
POTAMANTHIDAE	POTAMANTHUS Pictet, 1845	*luteus* (Linnaeus, 1758)	-
SIPHLONURIDAE	SIPHLONURUS Eaton, 1868	*armatus* Eaton, 1870	Large Summer
		lacustris Eaton, 1870	Large Summer
		alternatus (Say, 1824)	Large Summer

Appendix B

References

a. Freshwater Biological Association's Scientific Publications:

Edgington, J. M. And Hildrew, A. G. (1995). *Caseless Caddis Larvae of the British Isles.* Publication No. 53.

Elliott, J. M., and Humpesch, U. H. (1983). *A Key to the Adults of the British Ephemeroptera.* Publication No. 47.

Elliott, J. M., Humpesch, U. H. and Macan, T. T. (1988). *Larvae of the British Ephemeroptera.* Publication No. 49.

Kimmins, D.E. (1972). *A Revised Key to the Adults of the British Species of Ephemeroptera, with notes on their Ecology.* Publication No. 15.

Macan, T. T. (1970). *A Key to the Nymphs of the British Species of Ephemeroptera, with notes on their Ecology.* Publication No. 20.

Wallace, I. D., Wallace, B. And Philipson, G. N. (1990). *A Key to the Casebearing Caddis Larvae of Britain and Ireland.* Publication No. 51.

b. Field Studies Council publications:

Ian Wallace, *Simple Key to Caddis Larvae,* 2006. Field Studies Council, OPI05. £5.00

Peter Barnard and Emma Ross, *A guide to the adult caddisflies or sedge flies (Trichoptera).* Field Studies Council, Test version 2007, available from the Field Studies Council.

Craig Macadam and Cyril Bennett, *A Pictorial Guide to British Ephemeroptera* Field Studies Council, Test version 2007.

David Pryce, Craig Macadam and Steve Brooks, *Guide to British Stonefly (Plecoptera) families: adults and larvae,* 2007. Field Studies Council, OP113.

The Riverfly Partnership, *River invertebrate monitoring for anglers.* 2007. Field Studies Council, OPI04.

c. Insects:

Harker, Janet (1989). *Mayflies.* Naturalists' Handbook 13, (Richmond).

Chinery, M (1993). *Field Guide to the Insects of Britain and Northern Europe.* Collins.

Goddard, J (1991). *Trout Flies of Britain and Europe.* A & C Black.

Miller, Peter (1995), *Dragonflies,* Natural History handbook no. 7, Richmond Publishing Co.

Southwood, T. R. E. and Leston, D. (1959), *Land and Water Bugs of the British Isles,* Warne.

Steve Brooks (Editor), Richard Lewington (Illustrator) (1997), *Field Guide to the Dragonflies and Damselflies of Great Britain and Ireland,* British Wildlife Publishing.

Appendix C

The amended DoE/NWC 'Biological Monitoring Working Party' (BMWP) score system for sensitivity to pollution (10 = high)

GROUP	FAMILIES	SCORE
Mayflies (or Upwinged flies)	Siphlonuridae, Heptageniidae, Leptophlebiidae, Ephemerellidae, Potamanthidae, Ephemeridae	10
Stoneflies	Taeniopterygidae, Leuctridae, Capniidae, Perlodidae, Perlidae, Chloroperlidae	10
River bug	Aphelocheiridae	10
Caddis or Sedge flies	Phryganeidae, Molannidae, Beraeidae, Odontoceridae, Leptoceridae, Goeridae, Lepidostomatidae, Brachycentridae, Sericostomatidae	10
Crayfish	Astacidae	8
Dragonflies	Lestidae, Agriidae, Gomphidae, Cordulegasteridae, Aeshnidae, Corduliidae, Libellulidae	8
Mayflies (or Upwinged flies)	Caenidae	7
Stoneflies	Nemouridae	7
Caddis or Sedge flies	Rhyacophilidae, Polycontropodidae, Limnephilidae	7
Snails	Neritidae, Viviparidae, Ancylidae	6
Caddis or Sedge flies	Hydroptilidae	6
Mussels	Unionidae	6
Shrimps	Corophiidae, Gammaridae	6
Dragonflies	Platycnemididae, Coenagriidae	6
Bugs	Mesoveliidae, Hydrometridae, Gerridae, Nepidae, Naucoridae, Notonectidae, Pleidae, Corixidae	5
Beetles	Haliplidae, Hygrobiidae, Dytiscidae, Gyrinidae, Hydrophilidae, Clambidae, Helodidae, Dryopidae, Elmidae, Chrysomelidae, Curculionidae	5
Caddis or Sedge flies	Hydropsychidae	5
Craneflies/Blackflies	Tipulidae, Simuliidae	5
Flatworms	Planariidae, Dendrocoelidae	5
May flies (or Upwinged flies)	Baetidae	4
Alderflies	Sialidae	4
Leeches	Piscicolidae	4
Snails	Valvatidae, Hydrobiidae, Lymnaeidae, Physidae, Planorbidae	3
Cockles	Sphaeriidae	3
Leeches	Glossiphoniidae, Hirudidae, Erpobdellidae	3
Hog louse	Asellidae	3
Midges	Chironomidae	2
Worms	Oligochaeta (whole class)	1

Appendix D
Simplified invertebrate monitoring for anglers

How to sample

The sampling method is similar to that used by the Environment Agency; a kick/sweep sample using a standard pond (hand) net for 3 minutes. This allows for comparable samples to be taken at a later date.

Kick sample using a standard pond net

The different habitats within the sampling area should be identified. For example, fast moving (riffles), slow water (pools), bank side (tree roots etc) and weed-beds.

A total sampling time of three minutes is split proportionally between the number of habitats identified for sampling. So, for example, if riffles occupy 50% of the site they should be sampled for 1½ minutes and if weeds occupy 25% of the site they should be sampled for 45 seconds.

The pond net is rested on the river bed and the area immediately upstream disturbed ("kicked") so that invertebrates are carried into the net by the current; further samples are taken moving across and/or upstream. For sampling in weed areas the net is "swept" through the weed-bed.

"Washing" the sample:

To ease the counting process one should remove as much of the unwanted debris from the sample as possible without losing any of the required invertebrates. Tip the contents of the net (3 minute kick/sweep sample) into a large bucket of river water. The water is then "strained" back through the net whilst agitating the stones and gravel to dislodge all the invertebrates. This will leave behind the unwanted heavy material of the sample. The bucket is then refilled with fresh river water and the process repeated until all of the invertebrates have been dislodged from the stones and gravel in the bucket. After a quick check is made that no invertebrates remain (such as heavy-cased caddis larvae), the stones and gravel can be discarded from the bucket. (Note: you might collect a number of empty caddis cases which should be ignored).

Holding the net into the current and moving the material around in the net will then remove most of the unwanted fine silt through the mesh. Some of the plant material can be removed after ensuring that any invertebrates have been dislodged. The remaining sample can then be returned to the bucket, half filled with clean water and the sample is ready to be sorted.

Target groups are transferred into a divided tray ready for counting

Sorting the sample:

The invertebrates should now be much easier to find providing that only small "sub-samples" are taken from the bucket at a time (using a small aquarium net) and placed into a shallow white tray half filled with clean water. If a large enough tray is used all (or most) of the sample can be processed in one go. The required invertebrates can then easily be picked out using a pipette (turkey baster) and transferred into some form of divided tray ready for counting.

Targeted taxonomic core groups:
The core groups for monitoring are:

Nymphs of stoneflies – Plecoptera
Larvae of cased caddis – Trichoptera
Larvae of caseless caddis – Trichoptera
Nymphs of up-winged flies – Baetidae (olives)
Nymphs of up-winged flies – Heptageniidae (Flat-bodied stone-clingers, e.g. Yellow May dun)
Nymphs of up-winged flies – Ephemeridae (Mayfly)
Nymphs of up-winged flies – Ephemerellidae (Blue winged Olive)
Gammarus – Freshwater Shrimp

(It may be beneficial to include other groups locally).

Recording Data:
Data can then be recorded on structured sheets (copy attached), using the following abundance estimates:

1 – 9 = A
10 – 99 = B
100 – 999 = C
over 1000 = D

Records should ideally be held on some form of database by the monitoring organisation.

Registering sampling site with the Environment Agency :
The representative of the monitoring group must qualify by having attended one of the Riverfly Partnership/John Spedan Lewis Trust for the Advancement of the Natural Sciences courses, and must discuss and register the sampling site with the Ecological Appraisal Team Leader of the regional Environment Agency office, noting the:

- name of group/organisation carrying out the sampling.
- name of the river.
- 8-figure grid reference of proposed sampling site.
- regularity with which sampling is planned e.g. monthly. NB. Greater regularity may be needed to pick up pollution events, such as from sheep dip.

The Environment Agency representative will:

- ensure the suitability of the site (avoiding conflict with established monitoring sites).
- provide information of data for the locality and help indicate a seasonal baseline for that site's "trigger" points. These are the points at which the count of

invertebrates is so low that the EA agrees to take action as they may indicate a potential serious problem. As stability and credibility of data builds, the trigger points can be revised.

- confirm contact details and action to be taken on trigger points being reached.

Equipment list:

Health and safety recommendations, e.g. reference to disease, beware of urban debris.
Safety equipment, e.g. life jacket.
Waders.
Standard hand net (commonly called 'pond net'). Standard frame plus a 1mm net.
Bucket.
Large white tray.
Small aquarium net.
Pipette (turkey baster).
Hand lens (for interest).
Recording materials (i.e. recording chart and guide for method and identification).

Appendix E Times of emergence of the Ephemeroptera

Species	Common Name	Peak flight months (weather dependent)											
3 Tails/Large Hindwings		J	F	M	A	M	J	J	A	S	O	N	D
Ephemera danica	Mayfly				✓	✓	✓	✓	✓	✓	✓	✓	
Ephemera vulgata	Mayfly					✓	✓	✓	✓				
Potamanthus luteus	-					✓	✓	✓					
Leptophlebia marginata	Sepia dun				✓	✓	✓						
Leptophlebia vespertina	Claret dun				✓	✓	✓	✓	✓				
Paraleptophlebia cincta	Purple dun					✓	✓	✓	✓				
Paraleptophlebia submarginata	Turkey Brown				✓	✓	✓	✓					
Habrophlebia fusca	Ditch dun.					✓	✓	✓	✓	✓			
Serratella ignita	BWO				✓	✓	✓	✓	✓	✓			
Ephemerella notata	Yellow Evening Dun					✓	✓						
3 Tails/No Hindwings													
Brachycercus harrisellus	Angler's Curse							✓					
Caenis luctuosa	Angler's Curse						✓	✓	✓	✓			
Caenis macrura	Angler's Curse					✓	✓	✓	✓				
Caenis pseudorivulorum	Angler's Curse						✓	✓	✓	✓	✓		
Caenis rivulorum	Angler's Curse					✓	✓	✓	✓	✓			
Caenis robusta	Angler's Curse						✓	✓					
2 Tails/No Hindwings													
Cloeon dipterum	Pond Olive					✓	✓	✓	✓	✓	✓		
Cloeon simile	Lake Olive			✓	✓	✓	✓	✓	✓	✓	✓	✓	
Procloeon bifidum	Pale Evening dun				✓	✓	✓	✓	✓	✓	✓		
2 Tails/Large Hindwings													
Siphlonurus armatus	Large Summer dun					✓	✓	✓	✓				
Siphlonurus lacustris	Large Summer dun					✓	✓	✓	✓				
Siphlonurus alternatus	Large Summer dun					✓	✓	✓	✓				
Ameletus inopinatus	Large Summer dun					✓	✓	✓	✓				
Ecdyonurus dispar	Autumn dun						✓	✓	✓	✓	✓		
Ecdyonurus insignis	Large Green dun					✓	✓	✓	✓	✓	✓		
Ecdyonurus torrentis	Large Brook dun			✓	✓	✓	✓	✓	✓	✓			
Ecdyonurus venosus	Late March Brown				✓	✓	✓	✓	✓	✓	✓		
Electrogena affinis	None							✓	✓				
Electrogena lateralis	Dark dun					✓	✓	✓	✓	✓			
Heptagenia suiphurea	Yellow May dun					✓	✓	✓	✓	✓	✓		
Kageronia fuscogrisea	Brown May dun					✓	✓						
Rhithrogena germanica	March Brown			✓	✓	✓							
Rhithrogena semicolorata	Olive Upright				✓	✓	✓	✓	✓	✓			
2 Tails/Small Hindwings													
Alainites muticus	Iron Blue dun				✓	✓	✓	✓	✓	✓	✓		
Baetis fuscatus	Pale Watery dun					✓	✓	✓	✓	✓	✓		
Baetis rhodani	Large Dark Olive			✓	✓	✓	✓	✓	✓	✓	✓	✓	
Baetis scambus	Small Dark Olive		✓	✓	✓	✓	✓	✓	✓	✓	✓	✓	
Baetis vernus	Medium Olive				✓	✓	✓	✓	✓	✓	✓		
Nigrobaetis niger	Iron Blue dun				✓	✓	✓	✓	✓	✓	✓		
2 Tails/Tiny Spur Hindwings													
Centroptilum luteolum	Small Spurwing				✓	✓	✓	✓	✓	✓	✓	✓	
Procloeon pennulatum	Large Spurwing					✓	✓	✓	✓	✓	✓		

95 Kingfisher *Photo: Dennis Bright*

Chapter Five

CHALK STREAM FISH AND THEIR MANAGEMENT

The fish community of the chalk stream has been shaped through millennia by nature and, increasingly significantly in recent years, by man. Some species, for example the brown trout, capable of living in both salt and fresh water, moved inland from the sea at the end of the last Ice Age, around 10,000 years ago. Others, such as the pike, apparently inhabited southern English rivers at a time when Britain was connected by land to northern France and our rivers were mere tributaries of the huge river systems of continental Europe, before the seas rose to form the English Channel some 3000 years ago. Latterly, however, man has restructured chalk stream fish stocks through land-use systems in river catchments, pollution, water regulation and, not least, fishery management practices such as the introduction of non-native species like rainbow trout and non-local species like grayling.

This chapter addresses two major areas, dwelling upon the biology of some of the chalk stream fish species before consideration of aspects of management of both wild and artificially-reared stock. Descriptions of the fishes' life-cycles are of necessity fairly brief and selective, but are of the utmost importance as a basis for management of chalk stream fish stocks.

The fish of the chalk streams

Brown Trout, *Salmo trutta* (Linnaeus 1758)

The most popular sport fish of the chalk stream is the brown trout, an enormously adaptable and diverse species, some fish spending their entire lives in freshwater, others spending time between freshwater and river estuaries ('slob' trout) whilst others migrate fully to the open sea to feed, returning to rivers to spawn (sea trout). For the brown trout in the chalk streams, life begins in a gravel nest (the redd) where the orange eggs, 3-5 mm in diameter, are deposited by the hen fish, commonly in November to January. Stocked fish, selected from early spawning 'strains', tend to create redds in November whilst the native, wild fish will often not spawn until Christmas or even early into the New Year. A hen fish will lay 1000-1500 eggs per kilogram of her bodyweight, incubated in the gravel in well-oxygenated, cool water; developmental rates are temperature-dependent, hatching occurring after about 40 days at 10°C, 80 days at 5°C. The hatchling, called an alevin, develops hidden in the gravel, living off its nutritious yolk sac for about a further month before emerging from the gravel as a small fish, some 2.5 cm long. Filling the swimbladder and finding food and cover are major priorities, all three needs usually satisfied in quiet refuges where flow is least, at the stream margin. Juvenile trout are highly territorial, dominant fish occupying the prime lies. The number of young fish any particular part of a river can support is governed by the quality of available habitat; the fish need enough space in the river (their 'territory') to find food and to feel removed from neighbours – territory size decreases where there are plenty of invertebrates for food, and rocks, stones or weed to separate would-be, aggressive neighbours. Our productive, southern chalk streams help the young trout grow very quickly; by the end of their first summer, the parr may be in excess of 10 cm and over 15 cm in length by their first 'birthday' (i.e. around the New Year); two-year olds will be about 20 cm and three-year olds around 30 cm. In contrast, a brown trout from a cold, acidic Scottish stream may be half this length at three years!

Usually during their second spring of life, some of the trout parr (15-20 cm in length) will metamorphose to silvery sea trout smolts and migrate to sea. Not all rivers contain stocks of sea trout and quite what it is that makes an individual trout migrate or not is a point of debate; present thinking suggests a combination of genetic and environmental factors, such as food availability. The majority of sea trout are female, perhaps driven by reproductive energy requirements which can only be satisfied at sea. Some of these migrating trout will meander back and forth between fresh and sea water: these are the so-called 'slob' trout. Others will move away to sea but return to fresh water after only a few months away as 'finnock' of 25-35 cm. Yet others may spend a winter or two at sea before returning to fresh water as sizeable sea trout of 1-2 kg, exceptionally in excess of 5 kg. Sea trout, unlike salmon, *do* feed in freshwater and do not suffer such massive rates of post-spawning mortality. Studies in Scotland have shown individual sea trout spawning up to seven times in successive years. Southern English rivers contain important stocks of sea trout, not only in the well-known fisheries on the lower Test and Itchen but also in some of the tiny New Forest spate streams.

96 Trout parr

Sexual maturity for brown trout is reached at 2-3 years of age with life expectancy of 5-8, exceptionally 10 years. Fish ready to spawn may migrate considerable distances upstream within a river if suitable sites are not close-to-hand. Sexually mature males are easily distinguished (even in water) by the darkened body and 'pointed' head with its hooked lower jaw (the 'kype'); ripe females tend not to darken as extremely as the males, and out of water display a soft, distended belly, supporting an egg mass which may be more than 20% of her bodyweight.

Many of our chalk stream reaches may have been stocked at some time in the last 200 years or so or are still stocked today. The Environment Agency's National Trout and Grayling Fisheries Strategy has attempted to control areas where stocking should or should not occur to conserve locally native stocks of wild brown trout (for example, on the upper reaches of the Itchen). The scientific literature remains rather equivocal on the effects of stocking on wild brown trout – there is for sure much literature that describes problems such as genetic contamination of local, wild strains by introduced stock or displacement of wild fish by stockies; many scientists, owners, and keepers support this view. But there is a body of other work, including recent research by the Game Conservancy Trust, which suggests that wild fish, adapted to life in their own part of the river, cope well with introduced aliens. The GCT work indicates that stocked fish do not grow well in the river and are prone to much wandering (usually downstream and up to 9 km from the point of stocking), apparently muscled out by the locals. The research did confirm however, that some stocked diploids did interact with wild stocks during spawning, increasing the potential for introducing domestic behavioural traits into the wild component of the stock. However, whilst the scientific literature does remain equivocal and it seems that potential effects from stocking vary between locations, the 'precautionary principle' *against* stocking in some areas is surely wise in an attempt to support wild brown trout. It makes little sense to stock fish where the habitat is rich and varied, where wild fish can thrive *and* where anglers' expectations can be managed so that the wild stocks produce acceptable, sustainable sport in the absence of numerous, usually larger stocked fish. In contrast, it also makes little sense to attempt to promote wild stocks where the habitat is all wrong, for example in the deep, wide middle reaches of the Test or Itchen, or perhaps where commercial realities dictate angling pressure in excess of what a wild stock alone could sustain.

It is important for anglers to be able to differentiate stocked from wild fish. (97, 98) show two contrasting fish, one a really good quality, stocked fish, the other a true wild fish with sleek body, perfect fins and bold, haloed, red spots.

97 A good quality stocked brown trout.

98 A wild brown trout.

Atlantic Salmon, *Salmo salar* (Linnaeus 1758)

Like the sea trout, the Atlantic salmon grows to adulthood at sea but spawns, and spends its juvenile phase, in freshwater. The salmon is also a winter spawner (November-February) cutting redds significantly deeper and larger than brown trout, anything up to several metres in length and more than 30 cm deep. Riffles of clean gravel are chosen as spawning sites, no different from the brown trout. Indeed, one keeper (at least?) on the Itchen used to fence off the trout redds to prevent the later-spawning salmon 'over-cutting' and disturbing the buried trout eggs!

99 Salmon: four stages of development

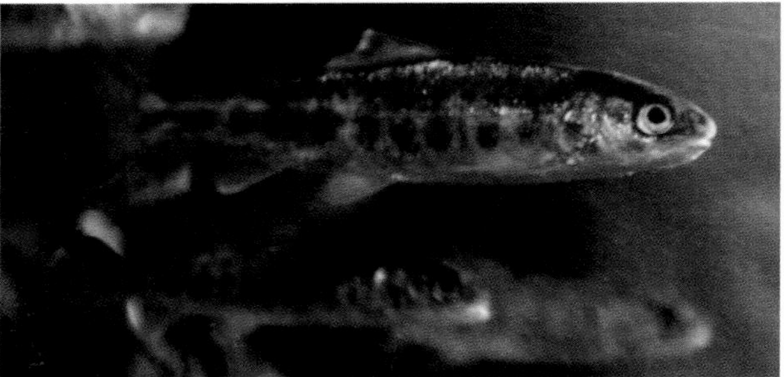

100 Salmon parr

The hen salmon produces around 1000 eggs per kilogramme of her bodyweight, each 4-7 mm in diameter. These hatch in 45 days at 10°C, the alevins developing in the gravel for another month or so. In the hatchery, 80% of fertilised eggs may be expected to survive to swim-up fry; in the river, 60-90% may be possible, although chalk stream survival rates are often much lower, not exceeding 20%. Scientific work on the Itchen during the 1980's indicated emergence rates of alevins from the gravel of 0-17% of eggs laid, the result of siltation and concretion of the gravel, suffocating the incubating eggs. These kinds of hatch rates will not support a salmon population because subsequent survival of the young fish is naturally and invariably low as they succumb to starvation and predation; maybe not more than 10% of the youngsters will survive their first year in the river. Salmon fry and parr most commonly inhabit shallow riffles, sheltering in aquatic weed, sharing a similar habitat to brown trout juveniles.

A Salmon (near smoulting)	B Brown trout parr
• **Salmon tend to have a narrow head with a bluntly pointed snout** • **Mouth (maxilla) – does not extend back of pupil** • **Tail fin – pointed lobes and deep fork in salmon** • Gill cover (operculum) spot – *usually* one single black spot on gill cover of salmon • Pectoral fins – proportionally larger in salmon than in trout • Parr marks – very pronounced in salmon, less well pronounced in trout • Black spots – rarely below lateral line in salmon often below lateral line in trout	• **Trout tend to have a wider head with a more rounded snout** • **Mouth (maxilla) – extends beyond back of pupil in trout** • **Tail fin - rounded lobes and shallow fork in trout** • Pelvic and anal fins – usually white leading edge on these fins in trout but not in salmon • Adipose fin – *usually* red in trout but not in salmon • Red spots – usually only along lateral line in salmon, more widespread in trout • Spot halos – salmon spots do not have white-ish halos, trout spots usually do

101 Juvenile salmon and trout, showing their obvious differences.

Young salmon and trout are easily confused; (101) indicates some of the differences, including the long, elegant pectoral fins of the salmon, its more forked tail and slender body shape, particularly towards the tail. When the salmon parr reach a length of around 15 cm, usually in 1 year in the southern chalk streams, they undergo a metamorphosis in readiness for life at sea – most obviously, the flanks silvering as the fish become smolts. In the highly productive environment of the River Test, for example, greater than 99% of salmon parr become smolts after a little over a year in freshwater, leaving the rivers in April or May to head off to their feeding grounds in the North Atlantic (contrast this with the situation further north, say Iceland, where 1 year-old smolts are not seen, the fish taking usually 3 or 4 years to grow to smolt). The number of smolts leaving our rivers each year is incredibly variable, based on counts taken by the Environment Agency at Nursling on the Test and Gater's Mill on the Itchen. In-river survival is influenced by a host of factors such as habitat availability, water quality, and predation. There is some evidence of significant predation on young chalk stream salmon by cormorants.

Those fish that do make it to the sea grow very rapidly on a nutritious diet of fish and crustaceans such that fish returning to freshwater after just one winter at sea (known as grilse) may be 50-65 cm long and 1.5-3.5 kg in weight. The vast bulk of chalk stream salmon nowadays are grilse (perhaps 80-90%+), returning to the rivers during the summer months. Multi-sea-winter salmon (fish of around 4kg plus) have become increasingly uncommon in the Test and Itchen in the last two decades, although during the 50's, 60's and 70's, these bigger fish usually comprised more than 70% of the returning stock. Survival rates from smolt to adult are unsurprisingly low, in the order of 1-5%, although estimates for the Itchen suggest rates up to 11%. Causes of mortality at sea include predation, commercial fishing and water quality, though it appears that the over-riding issue is one of climate-driven change in the marine environment disrupting food chains and thus the salmon's ancestral winter feeding grounds. So, a salmon smolt might make its way down the Test or the Itchen, swim 3000 kilometres up into the North Atlantic, avoiding predators and commercial fishermen, only to find its winter larder is largely empty. As if all this is not enough, those fish that do survive life at sea return home to hostile high temperatures in coastal waters, and low flows and very high summer river temperatures. Little wonder, then, that the salmon is in such big trouble in central southern England. Returning adult salmon enter our rivers anytime from early Spring to late Autumn, encouraged especially by flushes of fresh water coming down the river. Once in freshwater, they do not normally feed, but move upstream in fits-and-starts to find suitable spawning sites, resting periodically in sheltered lies, protected from flow and light. As cocks and hens pair up on the gravels, the cycle is nearly complete; post-spawning mortality rates are huge (greater than 90%), although a very few spent fish ('kelts') may repair and survive to spawn again another year. Scales from a giant Welsh hen salmon in the author's possession indicate it had apparently spawned five times – a real survivor.

Rainbow Trout, *Oncorhynchus mykiss* (Walbaum 1792)
The rainbow trout is native to the west coast of North America, inhabiting the rivers draining into the Pacific from northern Mexico in the south up to Canada in the north. The species was introduced to the British Isles during the 1880s and has been spread to all areas of these islands as a fish farmed for food and sport. Very few (perhaps as few as six) self-sustaining populations have been established in the UK (none known on the Test or Itchen) although spawning does periodically occur in many more locations. Thus, most stocks in Britain are the product of very recently introduced farm-reared fish; their life expectancy in the river can be measured in weeks, days, even hours. The fish has a life-cycle similar to the brown trout, the female laying 1500-2500 eggs per kilogramme of her bodyweight, 3-5 mm in diameter, in a redd in well-oxygenated gravel. Spawning occurs from October onwards, although perhaps most commonly where it occurs in the British Isles, from February to March. This tendency to late spawning has been offered as a contributory factor to the limited number of natural spawning sites in Britain; it is suggested that brown trout fry, hatching earlier, out-compete the less advanced rainbow trout fry. Also, most wild-spawning rainbow stocks are found in spring-fed, alkaline waters with fairly stable temperature profiles. However, why the fish has not become more widely naturalised remains a perhaps fortuitous mystery. Naturally-hatched juvenile rainbow trout are occasionally encountered in the chalk streams, bearing 5-10 parr marks and numbers of small black spots along silvery flanks; these young fish are usually more stout in the body than brown trout or salmon of a similar length (102).

The diet of rainbow trout is similar to that of the brown trout, featuring a range of aquatic and terrestrial insects and small crustaceans and molluscs. Large fish may also predate on smaller fish of all species. Growth rates of rainbow trout are far greater than

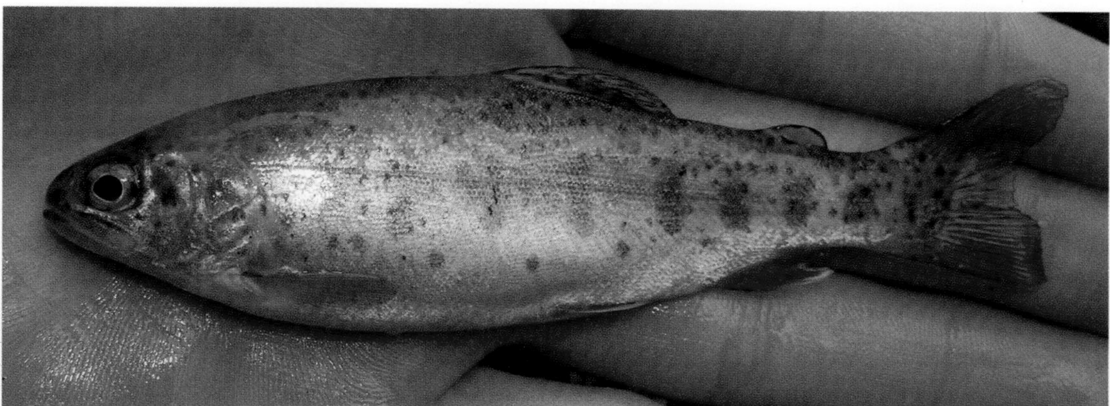

102 A wild British rainbow trout parr.

those of brown trout, with farmed fish capable of reaching a kilogramme in 12 months and 2 kg in 18 months. Farmers are producing ever-larger fish, to date in excess of 15 kg. Because of its lower cost and relative ease of capture by anglers, the rainbow trout is the principally stocked trout species in Britain as a whole (although not in the chalk streams). For the river keeper, the rainbow does give a chance to stock a fish that might take during the hot days of summer whilst the browns keep their heads down. Interestingly and contrary to much common perception, some chalk stream keepers regard large rainbows as a greater challenge to the rod than brown trout! The species is however, fairly short-lived, with a maximum age of 5-6 years in Britain.

Grayling, *Thymallus thymallus* (Linnaeus 1758)

In the angling world, the grayling has been something of an enigma – to some a 'coarse' fish, not worthy of pursuit, but increasingly to many a 'game' fish, the 'lady of the stream'. It was introduced to the Test early in the 19th century and to the Itchen not until the 20th century. The species is unmistakeable, characterized by a huge dorsal fin (especially in the males) and, when held, a unique smell (supposedly of thyme, thus the scientific name). In the water, grayling can be differentiated from trout by the large dorsal fin, if visible, but especially by the grey colouration, more slender rear end and deeply forked tail. Shoaling juvenile grayling often tend to swim in mid-water on gravel riffles, whereas trout will tend to hug the bottom for most of the time. The close relation between salmon and trout and the grayling is indicated by the presence in all three of the adipose fin, the small fleshy fin on the back. The reproductive biology, too, bears similarities as grayling spawn in shallow water laying 3-4 mm eggs in a gravel redd. Males guard and fight for prime spawning territory and display most spectacularly to a fancied mate. Spawning typically occurs later than salmon or trout, usually in March in the southern chalk streams. The eggs are incubated for about three weeks, producing a tiny alevin around 1 cm long. Once this juvenile has exhausted the food supplies of its yolk sac, it emerges from the gravel to feed on minute aquatic invertebrates such as midge larvae; as they grow, grayling take ever larger food items including shrimps, cased and uncased caddis larvae and various mayfly nymphs. In the stomach contents of brown trout and grayling on the upper Test. I have noticed a high degree of overlap, both sharing really abundant food items like shrimps and caddis larvae; scientific studies elsewhere suggest that the two species eat a subtly different diet. Both apparently also eat their own (and each others') eggs at their respective spawning times. Large grayling probably eat some small fish; in Scandinavia, big fish are taken on Mepps lures. Southern English chalk stream grayling grow really quickly, reaching 15 cm at the end of their first year and up to 30 cm after two years. Sexual maturity is reached at 2-3 years. Fish older than 5 years are not common on the Test or Itchen, although the species may live for ten years. Maximum size in Britain is around 55 cm, 1.5 kg, although any fish over 500 grams is getting large.

103 Grayling

Today the grayling is not the persecuted fish it once was on the chalk streams, even as recently as twenty years ago. Then and before, fish were electric fished or netted in the autumn in their hundreds and thousands, destined for the lime pit. Today, many fisheries value their grayling and benefit from offering high quality, winter grayling days to discerning, appreciative anglers.

Pike, *Esox lucius* (Linnaeus 1758)

The pike is one of Britain's largest freshwater fishes; most commonly it is the top predator in a lake or river. It is regarded as one of the indigenous species which survived in southern England during the last Ice Age, but was redistributed or introduced to much of Britain and Ireland in the Middle Ages, transported in barrels by horse and cart for food and sport. Introductions of pike to waters without previous stocks continue today, very often illegally.

Pike spawn between February and May (commonly March) when water temperatures range from 4°C-11°C (by legend, coinciding with the flowering of the first daffodil!). They migrate in rivers to find calm, weedy backwaters, ideal spawning territory. Males scarcely exceed 4 kg in weight, and 2 or 3 or more of these smaller fish may attend a single female in the act of spawning. Eggs and milt are released, largely at random, in vast quantities; females produce 10,000-20,000 eggs per kilogramme of bodyweight. The eggs, each 2-3 mm in diameter, are sticky and adhere to submerged vegetation. Hatching occurs usually around 15 days after fertilisation, producing a larva which attaches to vegetation by means of an adhesive "organ" on the head. About a week later, having used up its yolk sac, the 6-8 mm fry swims freely, feeding in the first instance on appropriately-sized invertebrates, though fish soon feature on their menu, including their own siblings, when they are around 3 cm in length. Southern English chalk stream pike may reach 30 cm at the end of their first year, 40 cm after 2 years and 50 cm after 3 years. Sexual maturity occurs at 2-4 years of age. In Britain, pike grow in excess of 20 kg, most commonly when preying predominantly on trout in large lakes or reservoirs. In highly-managed chalk stream trout fisheries, fish in excess of 10 kg are very rare. Pike eat around 3% of their bodyweight per day, meaning that a 2 kg pike eats around 22 kg of prey fish per year. *If* (however unlikely) this 2 kg 'jack' was eating nothing but stocked brown trout, he might be costing the fishery around £150 per year in stocked fish! However, pike will try and eat any animal (whether covered in scales, feathers or fur) that is not too big to swallow; they prefer prey 10-25% of their own weight, although large fish will take others as much as *half* their own weight! It is not uncommon whilst electric fishing to capture a 2 kg pike which, when gently encouraged, regurgitates a 700 gram trout. Conversely, in recently examined gut contents of over 30 pike, many over 50cm long, not one of them had a fish (or their remains) in its stomach longer than 12cm! However, very importantly, pike can be effective at some degree of self-regulation of population through cannibalism with as much as half of the mortality rate of 2-3 year old pike inflicted by their own kind. (104) shows a view down the throat of a 7kg pike – the dark shape in the middle of the picture is the tail of a smaller pike, around 1 kg in weight! The issue of whether or not to attempt to control pike numbers in a river trout fishery is discussed further on page 130.

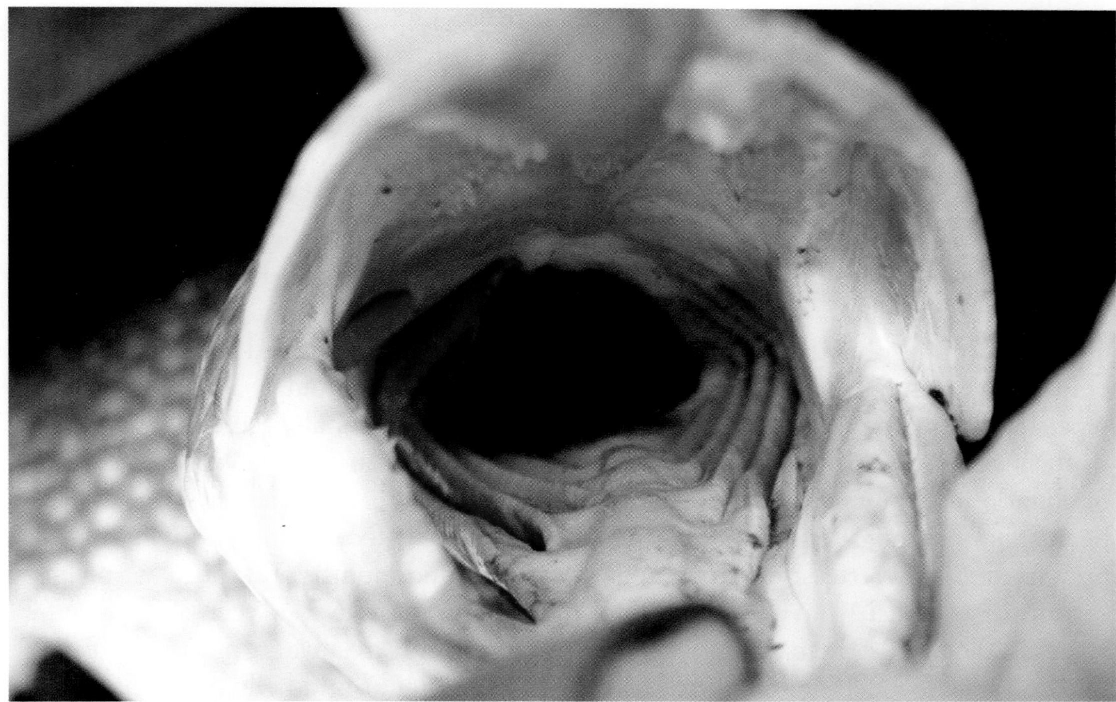

104 The view down the throat of a large pike; the dark shape in the middle of the photo is the tail of a smaller pike.

European Eel, *Anguilla anguilla* (Linnaeus 1758)

The European eel (105) displays perhaps the most amazing life-cycle of any British freshwater fish. Much of the detail of the life-cycle is supposed or pieced together from scant evidence; not much is known for certain.

105 European eels

Life apparently begins for the eel in the Sargasso Sea, east of the Bahamas. Here, each female lays millions of 1 mm eggs at depths of 100-200 m. The young eels, called leptocephalus larvae, follow the North Atlantic Drift back towards Europe. So unlike eels are these leaf-shaped larvae that when first found they were described as a totally new fish species, *Leptocephalus brevirostris*! Subsequently, it was shown how these odd creatures gradually metamorphose into minute 'glass' eels as they drift ever nearer the continental shelf of Europe. Their passage, north-eastwards across the Atlantic takes perhaps a year or more, and as the 6-7 cm long eels reach the coast, they develop pigmentation and become 'elvers'. The elver migration into freshwater in the spring is a similarly amazing voyage as the young eels negotiate all manner of obstacles, even crossing land to inhabit ponds and lakes. The author once watched a column of elvers crawl high up an outside wall of a fish hatchery building, attracted by a dripping overflow pipe! It is thought that females migrate further inland than males, generally stay longer in freshwater and grow larger. Eels have catholic tastes in food, eating a wide range of aquatic invertebrates and small fish, alive or dead. It is said that there are different head shapes amongst stocks of eels, dependent on their food items, those eating small prey (primarily invertebrates) having narrow heads and mouths and those eating larger organisms (like fish) having broad heads. Growth is slow: a four-year-old eel may be around 20 cm, 40-50 cm at 10 years of age. During this freshwater phase, the eels are known as 'yellow eels', alluding to the colour in the flanks and belly of the fish **(106)**. At some point in life, triggered by as yet unknown stimuli, yellow eels metamorphose, becoming silver in the flanks with enlarged eyes. These silver eels are now ready to return to sea to spawn, after 7-20 years (exceptionally 50 years) in freshwater. During this summer and autumn downstream migration the eels have historically been heavily harvested in licensed fyke nets and traps, providing income to keepers and owners along the chalk streams. Those that escape from the river navigate back to the Sargasso Sea, apparently guided by a combination of cues from the sun, stars, magnetic fields and an

106 A large 'yellow' eel.

107 Historic eel traps on the Test at Leckford *Photo: R H Dadswell*

acute sense of smell; after spawning, the adult eels die. No adults have yet been found in the Sargasso Sea, but the trail of their migrating babies points in that direction.

The status of eel stocks in Britain is of massive concern. Numbers are dwindling for a host of reasons not yet fully understood. One truly bizarre, probably devastating factor has been the recent introduction to the UK, including the Test and Itchen, of an exotic nematode worm parasite (*Angillicola crassus*) that infests the eel's swimbladder, quite possibly rendering the organ useless in providing buoyancy once the fish heads out to sea on its spawning migration. The fish may simply drift downwards into the ocean's abyss to a crushing death, unable to maintain a position in the upper layers. One consequence of this disastrous drop in eel stocks has been the cessation of trapping by many keepers, particularly as licence fees make the fishery only marginally worthwhile.

Other fish species

The chalk streams support a rich array of other fish species. There are a number of coarse fish species, some of which, like roach and chub, grow big in the Test and Itchen and are avidly sought-after as winter quarry, especially in the middle and lower reaches. Quite recently, carp and barbel have appeared in the chalk streams as, to many, unwelcome arrivals, escapees from fisheries and fish farms or possibly the result of deliberate introductions. It is likely, however, that they are here to stay. Carp in the middle and lower reaches of both rivers are quite established and growing big. For the barbel, the chalk streams may well make ideal habitat and they could grow very large indeed.

109 Bullhead

However, amongst our tiny, inconspicuous fishes, there are many amazing stories to be told. The bullhead or Miller's Thumb, for example, is normally invisible to the angler's eye (incredibly camouflaged and tucked-up under stones) but it is hugely important in the overall ecology of the stream. It could be, for example, that bullheads make up over a quarter of the total weight of all fish in the river! Whilst scientific study does not support the notion, they are widely regarded as trout egg eaters; Mick Lunn describes how his grandfather thought bullheads to be "gluttons for the spawn of fish, working the trout redds and gobbling up the eggs." Bullheads are, however, intriguing fish in their own right. In the spring, the male prepares a nest under a stone which he guards aggressively. His response to anything passing close by is to bite it: if it's food, he swallows it, if it's another male, he escorts it away, if it's a female, he pulls her into his nest, as a mate! Thus courted, she lays an egg mass, very often on the roof of the nest, i.e. the underside of large stones (108).

The male continues his guard duties over the eggs and hatchlings, though when their time comes, the young bullheads need to leave home quickly before they become a food item for their father. As they grow, bullheads eat invertebrates and other fish up to whatever size they can fit into their very large mouth. They in turn are important food items for a host of other river animals including trout, kingfishers and otters. Bullheads make excellent aquarium fish, quickly learning to take food from one's fingers, but they are now a heavily protected species, enjoying a conservation status similar to that of the salmon.

108 Bullhead eggs attached to the underside of a submerged stone

A similarly inconspicuous creature, for much of the year, is the brook or Planer's lamprey, one of three species of lamprey found in Britain, but the only one regularly seen on the Test and Itchen. These creatures are not fish at all but belong in a group of the most primitive vertebrate animals (the lampreys and hagfishes) whose design appears not to have altered significantly for 450 million years. Brook lampreys spawn on gravel in the spring in excavated depressions around 30 cm wide, up to a dozen animals writhing together in a tangled mass.

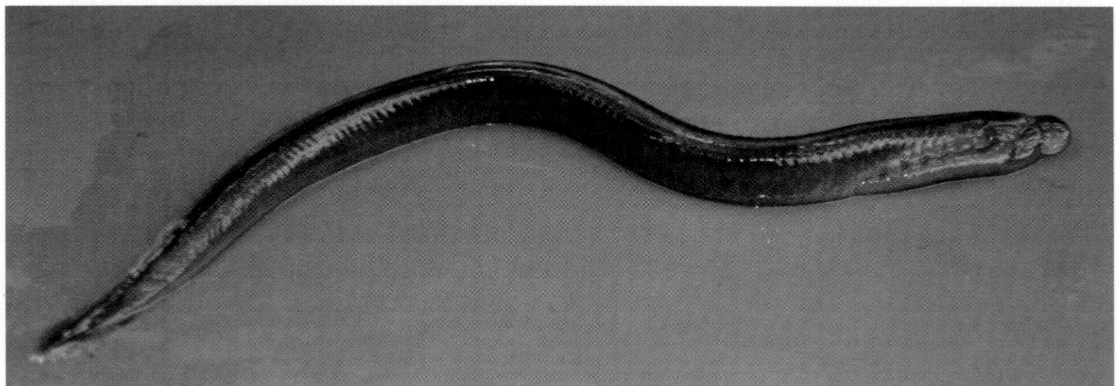

110 The ammocoete larva of the brook lamprey.

The young lampreys emerge from the gravel and drift downstream to settle into silted areas of the river where they live, for up to seven years, with only their snouts poking out of the silt, gathering microscopic food from the flowing water. These youngsters, technically called ammocoete larvae, are so unlike the adults that until the mid-19th century, they were thought to be a totally different species: they have no eyes, two strange flaps around the mouth and very undeveloped fins (110). When the time comes for the ammocoetes to mature, they develop a silvery coloration, big eyes and obvious, separate fins (111), emerge from the silt and move to gravel riffles to find mates; they do not feed as adults and in fact shrink in size during the transformation. Soon after spawning, the lampreys are spent and die.

111 An adult brook lamprey.

The Test and Itchen catchments are also inhabited by two alien fish species brought to the UK as ornamental varieties and which have in recent years attracted significant press coverage and scientific attention. The topmouth gudgeon (or clicker barb) only grows to around 10 cm and is a rather unremarkable looking fish, but it is of concern because it has the capacity to reproduce in vast numbers, to compete with and predate on other fish species and it may carry diseases problematical for some of our native coarse fish species. Interestingly, as one of its common names suggests, the fish makes clicking sounds when lifted from the water and handled. The other alien species, found in both the Test and Itchen catchments, is the sunbleak (or motherless minnow) which, like the topmouth gudgeon, is capable of reproducing quickly in massive numbers and again competition with native species is a major worry, even though a giant sunbleak is only around 7 cm in length. However, neither the topmouth gudgeon nor the sunbleak appear to do well in the rivers themselves but are happier in connected lakes and ponds.

Stock management

Productive rivers such as the Test and Itchen have, for many millennia, supported indigenous wild stocks of salmon and trout, although the 'natural' chalk stream habitat, a much-braided channel winding through tangles of willow and alder cover in the valley floor, is unlikely to support large numbers of fish. Nevertheless, the stock levels the rivers now support have been significantly affected in recent years by man's activities, with the river keepers, riparian owners, Environment Agency and CEFAS amongst others striving to combat the damage caused by land and river management practices, industry, urbanisation and apparent climate change. It is noteworthy that even where such external influences are minimal, salmon and trout populations will self-regulate through mechanisms such as competition for spawning sites and suitable habitats for juveniles and adults. In other words, a stream has an optimum number of fish it can support at any one time according to food and habitat availability. Natural reproduction or the introduction of excess stock above this 'carrying capacity' may not necessarily increase the yields from the fishery as the excess numbers of fish will be displaced (and probably perish) in the absence of food and/or good habitat; indeed, excess stocking may well cause long-term harm to any resident stocks or simply be a wasteful exercise.

Where the physical characteristics of the river allow, i.e. where suitable areas of river exist for all stages of the life cycle, and angling pressure is light, management practices can be directed towards maintaining the stocks as close as possible to the natural carrying capacity of the stream or possibly improving in-river conditions to *increase* the carrying capacity. However, where the river conditions are not suitable for all stages of the life cycle and/or angling pressure is more intensive than a wild stock alone can sustain, as is the case in many middle and lower reaches of the Test and Itchen for example, stocking of fish is required to maintain the fishery. The sections below address the practices which can be applied to attempt to optimise conditions for wild and stocked fish.

Management of wild stocks

It is convenient and logical to consider management practices for wild salmon and trout stocks by looking at the physical requirements of each stage of the life cycle and then working as far as possible to help supply those requirements. Inevitably and crucially, these practices cannot be considered in isolation; for example, it is of no use providing excellently prepared spawning sites (say by gravel cleaning) if inadequate suitable habitat is available for the resulting fry. Here, a 'habitat bottleneck' results because of the insufficient fry habitat available for the increased numbers of fish resulting from the gravel cleaning. The net result is that the excess numbers of fish above the carrying capacity of the reach will be displaced or, more likely, perish with no ultimate benefit to the fishery. So it is vital to take a holistic approach to the management of wild stocks, considering all parts of the life cycle, with the final target of the production of more large adult fish, the anglers' quarry.

Improving egg survival

Trout and salmon, as previously noted, lay eggs in gravel nests ('redds'). Typically, trout choose loose gravel, 1-2cm diameter, in water 20-40 cms deep flowing at 15-20 cms per second and ideally, apparently, where groundwater upwells through the river bed. Chalk streams suffer from a natural process of calcification of the gravels which tends to produce a compacted river bed; when deposited silt is mixed in, the resulting gravel is a poor medium for egg and alevin incubation. Female trout have been observed digging 'test' redds – if they sense the gravel is not right, they go elsewhere. The natural processes of gravel compaction have been hugely exacerbated in recent decades as land management practices, river engineering, abstraction and climate change have contributed to increased in-river siltation. Work by MAFF (now CEFAS) in the 1990's compared survival rates of salmon eggs in raked and unprepared gravels in the Itchen, in the most extreme example giving results of 61% survival in the former, 7.5% in the latter. Therefore, the river keeper can, during October or very early November, prepare the gravel by raking or digging with a garden fork or use portable water pumps (which can be borrowed from the Environment Agency) to loosen the gravel, remove accumulated silt and hopefully produce an environment in which the deposited trout or salmon eggs can develop bathed in a through-flow of clean oxygenated water (112).

Trout deposit eggs about 15 cms down into the gravel and large salmon perhaps twice this depth, so these should be our guidelines for gravel preparation. Spawning beds 3-4 metres long and 60-90 cms wide are ideal, with cover close by for the adults. Some keepers say that sites which have historically been used for spawning are best prepared since the fish may have a particular reason for depositing eggs in these areas (e.g. upwelling groundwater). In the absence of known spawning sites, consider the parameters of depth, flow and gravel size mentioned above in selecting an area for gravel cleaning. It is essential to look downstream of the intended prepared spawning sites, because the emerging fry will need a place to live: they will be looking for shallow areas with lots of marginal cover to tuck up out of the flow, away from predators, where food is readily available.

112 A cleaned patch of gravel shows up clearly against the uncleaned gravel surrounding it.

Artificial spawning beds can be created by importing ungraded gravel (roughly 1-5cm diameter) which tends to be less liable to silting-up and to being washed downstream in heavy water. This is deposited on the existing stream bed, especially in areas where historical river engineering has excavated an over-deep and/or over-wide channel. Thus, the imported gravel is being used to attempt to reinstate what was there before. At the downstream end of each bed a row of boulders may help to limit displacement of the gravel. I know of one imported gravel bed (on the Wey) comprising 120 tonnes of gravel in a bed about 20m long and as wide as the river channel (around 6m). This bed seems to have been a huge success, autumn electric fishing surveys showing juveniles of several species (trout, dace, barbel, chub) and established beds of starwort and *Ranunculus* in an otherwise pretty weedless reach. It is imperative to ensure that any imported gravel does not in any way impound the river upstream of the created bed, because this will slow the flow and cause increased siltation in this area.

Good cover very close to the spawning areas must be available for the spawning adult fish so that in the event of danger they can dive into hiding. This could be downstream deeper water (perhaps scoured out by the boulders supporting the gravel bed), marginal cover along one unfished, 'wild' bank or temporary cover provided by means of tethered floating logs close to the spawning beds. Adult salmon or trout will be very wary of spawning in areas in which their own security is compromised by a lack of nearby cover.

Yet another possibility for optimising egg and alevin survival is the use of incubators mounted in or beside the stream. Early versions of these incubators (called Pahari or Kashmir boxes) were made of fine mesh (2 mm diameter mesh size) in which fertilised eggs, stripped from local broodstock, were incubated to hatching. More modern in-stream incubators (still commonly called egg boxes) are located below weirs or sluices to provide a head of water to feed the gravel-filled boxes in which the fertilised eggs develop (113, 114). Regular inspection is essential to clean off accumulated debris; then, once their yolk sac has been used up, the trout or salmon fry emerge from the gravel in the egg box and either flow out of the box to spread naturally or are gathered up for redistribution to areas of suitable habitat (see below). Artificial incubation systems are intended to maximise survival rates of eggs and alevins but are labour-intensive and prone to blockage or destruction by debris in the river and to vandalism. However, some use egg boxes to great effect; the Environment Agency, for example, gain 90% hatch rates from their box-incubated salmon eggs on the Itchen.

113 In-stream incubators or 'egg boxes' can provide a safe hatching environment for trout and salmon eggs.

Improving fry survival

As soon as they wriggle free of the gravel, tiny trout and salmon fry are washed downstream and hopefully to the margins, out of the main flow. Here they find shelter and easy access to food-rich marginal and riffle areas. The most ideal cover is provided by submerged stones or aquatic plants like Water-crowfoot (*Ranunculus*) or Starwort (*Callitriche*) or by marginal plants such as the sedges (*Carex* spp) which grow in a few inches of water but provide a dense, quiet 'forest' through which the young trout can roam. These young fish are generally regarded as intensely territorial, requiring concealment from one another as well as the many predators keen to eat them. In the right areas of the river, tiny solitary trout can be found hiding in the margins in spring; however, paradoxically, it is not uncommon to find 'families' of 2-6 fish occupying the same few inches of shelter. This marginal cover of plants will naturally develop where banks are stable, erosion from the river's flow is limited, livestock is fenced out and anglers' trampling is light or non-existent. Some artificial bank revetment methods such as corrugated metal sheeting or geotextiles (such as "Nicospan"), which are not recommended, produce sheer bank

114 Newly hatched trout

faces and restrict encroachment of marginal plants and therefore some fry cover is lost. In such cases, it is possible to create temporary, artificial cover downstream of spawning areas by anchoring tree branches in the margins; drifting debris catches on the twiggy branches, consolidating the slacks and gentle eddies within the cover, perfect shelter for tiny fish. Otherwise, the use of 'soft engineering' like faggoting and live willow bank facings as repair methods creates the sort of complex bank-edge ideal for little trout or salmon, especially if back-filling behind the bank-edge allows marginal plants to grow and hang over into the water to create lots of cover for the fish.

After a month or so in the margins of the chalk stream the 3-5 cm long fry spread out into the shallows of the main river, capable of swimming in the faster water but seeking shelter by weed beds and rocks on the river bed. It is possible in chalk streams to increase fry habitat by introducing flints into these shallows so that individual fry might occupy and defend a smaller territory, out of the view of neighbours; effectively, more fish can fit into the available space since each has a smaller but hopefully acceptable patch (115). Nursery areas for brown trout and salmon are definable in terms of their physical characteristics, particularly gravel and flint rich and ideally weedy riffles which are mostly found in the upper regions of a river or in small tributaries. This is certainly

115 A large flint is placed on a newly restored spawning riffle to provide emergent trout fry a place of refuge.
Photo: Dennis Bright

one of the main reasons why some of the middle and lower reaches of the Test and Itchen, deep and gentle flowing, sometimes canalised by historical engineering, are less suitable for salmonid spawning and early life rearing.

From parr to adult

As they grow bigger, brown trout and salmon parr seek progressively deeper water and the fishery manager or keeper can protect and promote stocks by providing cover, good conditions for suitable food organisms and by controlling numbers of competitors and predators (including anglers!). In the chalk streams, with their highly productive conditions, cover is usually available in the form of weed. Good management of weed beds ensures that both fish and their invertebrate food organisms have adequate cover though the low flows of recent years have created big problems with poor weed growth for many fisheries along both rivers. Some very simple methods can be applied where weed growth is sparse; for example, thinning dense overgrowing trees allowing light into the usually clear chalk stream is a logical start point. Trees are best coppiced or pollarded so that their roots remain to bind banks but they do not create a canopy over the river or its banks where helpful, marginal plants can grow. If beneficial weeds (like *Ranunculus*) are absent but river conditions favour their growth, planting is an option (see Chapter Six). Excessive tree clearance must be avoided, not least because the shade of a tree in itself provides a likely lie for a trout, though shaded areas can also provide important spawning areas and refuges from predation when river temperatures are high. In some areas, river conditions may conspire to limit weed growth, for example where the water is deep and the bed muddy. In such situations, the fishery manager can create further cover in a variety of simple ways, for example, through placing flints in the river which can provide holding areas for fish of all kinds, or logs or 'woody debris' can be anchored 15-20 cm off the stream bed to create all year round cover for adult trout. The Wild Trout Trust's *Wild Trout Survival Guide 2006* provides a good reference for carrying out such habitat enhancements (visit: www.wildtrout.org)

Many argue that the chalk streams continue to suffer from historical over-engineering. It is the case that the primary objective of the fishery manager should be to develop what is already available, for example, by encouraging where necessary the (usually) abundant chalk stream weed growth to provide cover for invertebrates and fish. In addition, he should look to work with nature and provide the ideal natural conditions that are vital for each stage of the fish's lifecycle. In particular he should provide good spawning habitat and cover in the winter, after the weed has died back and when fish are particularly vulnerable to predation.

Salmon fisheries

The salmon fisheries of the Test and Itchen, despite being a relatively short length of the river, have an important history. The famous salmon fishery on the Itchen at Woodmill was almost certainly yielding substantial catches during the reign of Edward III. It was not until 500 years later that these lower reaches became important sporting fisheries. In 1881, Broadlands, on the Test, took on its first river-keeper, John Cragg. In his first two seasons, 1881 and 1882, four rods killed 46 and 56 salmon respectively.

Now, the salmon stocks of the southern English rivers are at crisis point. The Environment Agency (and National Rivers Authority and Southern Water Authority before them) have monitored salmon runs through electronic fish counters, presenting a sad picture. For example, between 1990 and 2004, the number of salmon estimated to have entered the Test annually has ranged between 361 and 1155 fish, comparing almost unbelievably with 1950s and 1960s average *rod catches* of 900 fish per year in the Test, from a total estimated annual run of 4,000 fish. The potential reasons for this decline were discussed in the section on Atlantic Salmon, page 102.

From the point of view of managing the sport fishery for salmon, it must be recognized that the fish held in the lower fisheries are largely in transit, pausing before moving up to spawning grounds in the middle river. The holding period may vary from months to days or perhaps even hours. The free passage of salmon is vital and it is the responsibility of all who have charge of any hatches to ensure that these are manipulated with thought and consideration. Historically, there were tremendous rows because of the practice of some millers of closing their hatches overnight in order to develop maximum power from the river in the morning. This of course denied the salmon free passage and became an outlawed practice under the Salmon and Freshwater Fisheries Act of 1923 and its successors.

Although in more recent years spring salmon have been conspicuous by their absence, it is important to compare their behaviour with that of the smaller summer fish and grilse during their passage up river. Spring fish tend to be less hurried and show a marked preference for deeper lies and slacker water, particularly when water temperatures are high. Summer salmon and grilse on the other hand apparently seek areas of fast water, taking temporary respite in the shelter of boulders and weed beds. Therefore, the key to sound salmon fishery management is to provide suitable conditions which will attract and encourage the fish to take up lies during their upstream migration.

Some of the practices described in earlier sections have as much relevance in the salmon as the trout fishery. For example, the preparation of spawning gravels may be as beneficial to salmon as trout stocks and the creation and management of weed beds may provide good cover for either species. However, reviewed below are some of the techniques specifically favoured at present in the salmon fisheries for providing good lies for fish.

Buck pile or stake groynes or breakwaters provide a heavy scouring flow into an area of river that has a good gravel base, which should then remain clean, but also allow a pronounced back-eddy to form. These structures may jut out from the bank 2-3m and provide excellent holding spots close in, the fish tending to lie downstream of the outside edge of the groyne.

Many salmon fisheries regulate their anglers by limiting permitted methods. Most commonly, this involves restrictions on fishing 'natural' baits such as worms and prawns whereby these methods are disallowed completely or else for some part of the season, perhaps June onwards. This then leaves the angler with the choice of fly-fishing or spinning. Where catch-and-release is the norm, as on the Test and Itchen in these enlightened days, fly fishing, especially, is a favoured method of maximising survival rates of rod-caught fish. Least favoured is bait fishing since natural baits can lead to deep hooking.

As in the trout fishery, catch-and-release rules have a place in the modern salmon fishery. As previously noted, the Test and Itchen salmon fisheries have achieved 100% catch-and-release fishing against a national average just less than 50% (in 2005). Studies have shown that carefully handled fish have a very good chance of survival to spawning and thus of contributing to future salmon stocks.

2007 sees the official end of the Irish drift net fishery, thought by some to be responsible for taking significant numbers of fish destined for the southern English rivers. However, only time will tell whether the end of this fishery will lead to tangible improvements in our salmon stocks.

Again, however, the management of salmon stocks illustrates the absolute necessity for a broad-ranging perspective in fishery management so that the functioning of each facet of the river's ecosystem fits into a broader picture of the ecology of the entire catchment.

Regulation of fishing

Any fishery, whether wild or stocked, regulates its anglers in some way through limitations on, for example, catch-and-release fishing, method, length of season, bag or minimum takeable size, or number of rods allowed to fish at any one time.

In recent years, much debate has centred around the use of catch-and-release policies in salmon and trout fisheries. On the negative side, a very small minority of fish *will* die after release and, some say, those that are caught and returned become progressively more difficult to catch; there is also some evidence that catch-and-release policies, through boosting survival rates and thus overall fish numbers in the river, may actually depress growth rates in trout fisheries. However, studies and experience in Britain now show how valuable a tool catch-and-release can be in fish stock conservation since post-capture survival rates can be extremely high, for example over 80% for salmon. The Test and Itchen have been at the forefront of rivers where catch-and-release is used as a fishery management tool. In 1995, for example, none of the 167 reported salmon taken from the Test were returned but since 2001, only 5 of more than 1500 rod caught fish have been knowingly killed. This is an extraordinary achievement in the national context where annual catch-and-release rates, at present, are less than 50%. North American studies suggest that mortality rates for trout taken on artificial lures (including flies) may be less than 10%, compared to up to 50% for fish taken on natural baits which the trout are more prone to gulp down and thus be hooked deeply. In Dorset's River Piddle, survival rates in excess of 60% have been recorded for wild trout in a catch-and-release fishery at the end of the fishing season. Additionally, individual fish may be repeatedly caught so that such a limited stock is eked out; again, American studies show trout may be caught and returned alive up to six times in a season. However, the ideal catch-and-release rules of barbless hooks and release of fish in the water (i.e. the fish never actually leaves the water) suppose that the anglers are capable of executing these rules, or being helped to do so, in an effective manner.

Stocking fish

Even towards the turn of the 19th century, F.M. Halford recognised that many fisheries were unable to rely solely on their resident trout populations and required artificial stocking. This is particularly true, as mentioned previously, where river conditions for a given fishery do not favour natural reproduction and where a riparian owner needs to attract some significant return from his water. In these situations, stock reinforcement of some kind is an obvious option.

Species stocked

Brown Trout

The brown trout is the most popular trout species stocked in the chalk streams. It is synonymous with the chalk streams, anglers travelling from all corners of the earth for this fishing. Furthermore, the brown trout is native to the chalk streams and is therefore adapted to these conditions. If stocked brown trout are provided with good habitat, they generally wander less than rainbows within the fishing season. In river fisheries, catch returns of brown trout are good (i.e. the number of fish recorded as caught relative to the number stocked), with 70% a realistic target. However, over-wintering survival of stocked brown trout (and rainbows) is very low, so fish left in the river at the season's end are largely wasted to the fishery. One keeper (on the Loddon) electric fishes each November, retrieving over 600 stocked browns and rainbows, which he feeds and over-winters in his stews – he loses no more than a handful through the winter and then has valuable and well-mended fish with which to start the new season.

In recent years, a number of fish farms have developed 'triploid' brown trout for stocking. These fish are the product of a two-stage fish farming process which starts with the feeding of normal juveniles with food laced with synthetic testosterone, producing genetically normal male fish and others which retain their genetic 'femaleness' but acquire, from the feeding of testosterone, some degree of male sexual function – specifically, these genetic females grow testes and can be made to yield sperm. Incredibly, this sperm contains only female genetic material, so when it is used to fertilise eggs from normal females, the resulting eggs and young *must* be all-female. The second stage of the triploid-making process sees these all-female eggs treated with very high pressure or possibly a particular temperature regime causing abnormal development of their chromosomes, one product of which is that the fish are infertile. Triploid browns were originally developed to help fish farmers get their fish to stock size after the chemical malachite green was removed from the armoury of treatments against a killer fungal disease which strikes brown trout in winter, especially as they mature. Fishery managers and keepers then realised that triploids would have an additional benefit in that they could not contaminate the gene pool of any wild fish in the river. As a stock fish, triploids offer additional advantages in that they present a challenge to anglers throughout the season, even very late-on when diploid stock fish are sexually maturing. They may even gain weight through a season and more triploids are likely to survive over-winter and be in good condition for the start of the following season. Disadvantageously, triploids can be prone to wandering if faced with large numbers of diploid stocked fish or wild fish, they can be vulnerable to rapid changes in temperature and oxygen levels and they are currently more expensive to buy than normal, diploid brown trout. There appears little hard evidence for the notion that triploids prefer to shoal or that they predate heavily on trout and salmon parr. However, there is ongoing research work looking in more detail at the use of triploids in stocking programmes, due for publication in late 2007.

Rainbow trout

The stocking of rainbow trout seems more popular on the Test than the Itchen. Many fishery managers (and, by inference, their rods?) favour the sporting qualities of these fish. Production methods for rainbow are highly efficient, providing widely available stock (of very large sizes if required) at prices significantly less than brown trout. Some keepers regard large rainbows as extremely wily quarry. However, the species is very unlikely (beneficially) to breed in the wild and has a strong tendency to wander; they do not over-winter well and again any fish remaining in the river at the end of the season will probably be a loss to the fishery. It is suggested by some (and argued to the contrary by others!) that rainbows, often stocked as large fish, unsettle and displace the brown trout, being a more vigorous and errant species. The rainbow trout could be an additional option for stocking, perhaps 'sandwiched' between early and late introductions of brown trout. In this way, the rainbows are readily taken by the rods as the river's food supply peaks and hopefully limited numbers remain at the end of the season. Some keepers, for example, stock rainbow trout during the mayfly season (May/June) when trout of all kinds are easily taken by the rods. In summer, rainbows might keep the fishing going when temperatures soar and brown trout lie low in the water.

Stocking practices

For those fisheries that do stock, the question of the *number* of fish to stock into a river is influenced by the size of the fishery, the quality of the habitat that river section can offer the stockies and the number of anglers and their catch and kill rates. Suitable habitat is important for stocked fish as well as wild fish simply because, whilst stockies may live for no more than a season in the river, a lack of good cover will force them to wander very soon after stocking if they cannot find a good lie. The fish are looking for a spot in the river, protected on at least two sides, in a 'dead space' where minimal energy is required to hold station against the flow but where food is close by, passing, largely, on the 'invertebrate drift'. Turning trout out into a barren length of water with little shelter is largely pointless as the vast bulk of fish will simply emigrate elsewhere. Fortunately, the southern English chalk streams afford excellent shelter for much of their length in the copious weed beds, albeit that this cover does not develop fully until early summer. Some traditional keepering practices such as putting the river to bed at the season's end by cutting out all the weed and taking the fringes down very short may actually strip away much winter and spring cover for wild trout and any surviving stockies. Better, perhaps, to be more selective and leave untouched areas of both submerged and marginal weeds and to delay tidying up until the spring, before the fishing season starts.

The number of anglers fishing a water, and removing trout, fundamentally influences stocking, and re-stocking rates. *Maximum* rod pressure for one mile of good chalk stream should be no more than 2-3 rods per day. This avoids over-burdening of the fish and the banks of the river and is probably an acceptable level of occupancy to the fee-paying rod. In its crudest form, the stocking of trout, whether to a lake or river, is on a 'put and take' basis. It is efficient use of a resource that where trout are stocked for angling purposes, as many as possible are recaptured at some time. Over a season, it is realistic for a fishery to aim for a catch return of around 70% of the total number stocked; the rest are lost to predation, theft, disease or simply move elsewhere.

A crucial question at this stage is how many fish the river can support *at any one time*, because this will affect the efficiency of usage of the stocked fish and how often restocking is necessary. Published guidelines suggest stock levels of 60-250 two-year old fish per kilometre of river though certainly fisheries operate successfully at much less than these figures, with perhaps as few as 30 fish per kilometre. Only experience, perhaps through trial and error, results in reasonable stocking levels for a particular piece of water. For those coming new to a water with no access to its previous stock records, a good fish supplier should be an initial point of enquiry. His business may be to sell as many fish as possible, but he has seen and stocked a variety of waters and therefore will be able to provide some guidance. For future years, the keeping of a record-book is an excellent predictive tool as pictures emerge of where fish are being taken and during which months. Furthermore, calculation of catch/angler figures and percentage catch returns are additional indicators of the fishery's performance and may provide reasons for alterations to future years' stocking policies.

Restocking is carried out when the population has dwindled such that fishing becomes difficult. Many fisheries (usually lightly fished) stock a couple of weeks before the start of the season, then restock two or three times during the season. This may lead to variable fishing, particularly just before the next stocking is due, but is not a significant problem where angling pressure is light. As a rule of thumb, the more regular the stocking, the more consistent the fishing will be but the more costly stocking will be from increased delivery charges. Thus, on heavily fished waters, restocking may occur once or twice a month. The number of fish restocked roughly replaces catches although angling pressure through the year is unlikely to be consistent, so that obviously popular times, like the mayfly season, may lead to increases in the number of fish introduced. Again, though, stocking above what the particular fishery can hold is a waste of time, effort and money.

Fish for stocking may be moved around the river in a variety of ways when being distributed from a point of delivery by the fish farmer or from a club's own stew ponds. Aerated tanks of water may be carried on a suitable vehicle where banks are good, or in a punt. Introducing water from another catchment should be avoided as it brings a risk of introducing disease. Stocking boxes or cages may be dragged up or downstream to suitable release points, or alternatively, fish may be moved over short distances (no more than a couple of hundred metres) in plastic bins, carrying and releasing small numbers at any one point. Fish held in the dark during transportation for release will be significantly less stressed by the experience and, if covered over in bins or tanks, are less likely to jump out. Good distribution of stock is important, helping to ensure that the available lies are found by the fish and encouraging them to spread out after a highly gregarious life-style on the fish farm!

Size of fish to stock

The size of fish to be stocked is a matter of personal choice and finances, and of availability from the farms. Trout may be stocked as fry (also called 'young-of-the-year'), yearlings, or takeable-sized fish (usually 2-3 years old), or a combination of all three. The standard stocked brown trout is the 30-40 cm long fish, 600-750 gms in weight. There is an apparent pressure, presumably originating from the wishes of anglers, for ever-increasing sizes of stocked fish, so that some fisheries introduce an average fish well in excess of one kilo, with large numbers of fish twice this size. Where rainbow trout are introduced, these may be of similar sizes to the brown trout and commonly form 20-30% of the total number of fish stocked in a season.

The stocking of yearlings elicits a mixed response from fishery managers and keepers. Beneficially, the fish will be individually cheaper to purchase (in the very short term) and, if they survive to takeable size, could be difficult to differentiate from wild trout. On the other hand, yearlings rise freely to the fly and may therefore become a nuisance; survival rates for a further year are likely to be low (possibly 20-25%); and, finally, yearlings are prone to wandering, perhaps to someone else's water! Initial stocking levels of yearlings of 150-350 fish have been suggested per kilometre of river.

A further option for stocking is to introduce, between March and May, fry or parr of between 3-6 months old. These fish are relatively inexpensive to buy and, if any survive, will produce very 'wild' looking two or three year olds. However, mortality rates are massive – perhaps 90% will fail to see the following spring – and, therefore, very large numbers must be stocked, in the order of 2000-6000 per kilometre of river. If wild fish exist in the fishery and spawn and recruit successfully, the introduction of yet more fry might not be helpful. Whilst farmed fish are generally thought to be less 'fit' than wild fish, stocked fry may have grown faster by virtue of their early days on the farm and may therefore be bigger and stronger so they could displace the generally smaller, native fry. As discussed previously, scientific opinion is equivocal but rightly cautious on this matter. If fry or parr are stocked, they need to find the right sort of habitat: food-rich shallow riffles with lots of cover. Their introduction to unsuitable habitats, such as the deep glides of the lower reaches of a river, is unlikely to produce any worthwhile reward.

Important details in stocking

A few final thoughts on the details of stocking. It is important to tie stocking dates in with other fishery management operations, particularly weed cuts. Thus, the river is prepared to receive the fish, and disturbing weed cuts are not made where recently-introduced, unsettled fish have been stocked. Some river keepers feed trout pellets through the winter period to maintain the condition of over-wintering stock and hold the fish in their stretch of the river. This practice is generally not to be encouraged, as it can increase competition and predation pressures on juvenile wild stocks.

Lastly, the release of any fish, or its spawn, into inland water in England and Wales requires the permission of the Environment Agency under Section 30 of the Salmon and Freshwater Fisheries Act 1975 (in common parlance, the fishery manager seeking to stock fish approaches the Environment Agency for a 'Section 30 Consent'). The system is designed to allow the Environment Agency to monitor fish movements at a local and national level and hopefully reduce the spread of disease and unwanted species. As part of this process, it is strongly recommended that a valid health check is requested from the fish supplier. Where fish are being bought from a recognised, reputable supplier, it is normally straightforward and involves the completion of a simple form; some suppliers will take care of this administration on behalf of their customers. Filing of applications can be done on-line (through www.efishbusiness.co.uk) from April 2007.

Holding facilities

The ability of a club or private fishery to hold on-site, or even grown-on, numbers of trout for stocking offers a few obvious advantages such as greater flexibility in stock control and potentially reduced costs (where fish are bought small and grown-on to stock size). However, there are equally obvious *disadvantages* whereby holding facilities cost money to build, need somebody's time and attention (e.g. fish feeding, screen cleaning) and inevitably fish will be lost through theft, escape or disease. It is possible for a club to invest heavily in a holding unit, to stock it and to lose the *entire* stock in a disease outbreak or theft incident.

116 Raceways or 'stews' for growing on restocking trout. *Photo: Lawrence Talks*

Fish for stocking are usually produced in earth raceways or ponds (116). A typical earth rearing pond on a fishery would be approximately 10-20m long, 5-10 m wide and 1 m deep; raceways are narrower, perhaps 2-3m wide, 20-50 m in length with depth no greater than 1m. The banks of these narrower holding units are often reinforced, for example by corrugated metal sheeting, but the bed is gravel. A constant through-flow of water is essential; 100kg of trout require 100-200 litres of water per minute dependent on fish size and water quality and temperature. The fish can be held at 10-20kg per m^3 of water, so that a raceway 20m long, 2m wide and on average 1m deep, thus holding $40m^3$ of water, could be stocked with 400kg of trout (supplied with 400-800 litres per minute of water). This equates to about 700 standard, stock-size brown trout. These figures are obviously generalisations; decreasing flow and increasing temperature mean fewer fish can be held in a given pond or raceway. Fish are retained within the pond or raceway by screens which also serve to exclude unwanted fish species and flotsam and jetsam; regular cleaning of screens to maintain flow is vital and is obviously a major job during periods of weed-cutting or autumnal storms. The ideal holding unit is close to the keeper's or fishery manager's house, is partially or even completely drainable to allow for absolute stock control and maintenance work, and is secured from poachers or avian predators.

Feeding of fish is carried out by hand or automatic feeder. In a low-key, fishing club raceway system, a single daily feed early in the morning is adequate to maintain, not grow, the fish. The major fish food manufacturers employ some excellent field representatives (often former fish farmers) who may help with queries on fish husbandry in general.

Fish disease

Fish carry with them, at all times, organisms which can lead to problematical diseases. For example, a microscopic examination of the mucus of the skin of good, healthy fish will reveal low levels of several types of protozoan parasite; ordinarily, the fish and the parasite (at low levels) coexist at no great detriment to the host. However, disease problems arise when a fish is stressed and, with lowered resistance, is overwhelmed by a proliferation in the disease-causing organism. Common stresses arise from increased water temperature, overcrowding (for example during transportation) or spawning.

Diseases may be caused by multi-celled animals (e.g. fish leeches), single-celled protozoa (e.g. "white spot"), bacteria (e.g. *furunculosis*), viruses (e.g. IPN, Infectious Pancreatic Necrosis) and fungi (e.g. *Saprolegnia*). One or two conditions have no known causative agent (e.g. UDN, Ulcerative Dermal Necrosis), or are caused by unusual circumstances such as nutritional deficiency. A few fish inevitably develop congenital deformities which are untreatable. An example is curvature of the spine (scoliosis or lordosis) which results in a fish with a kink in its body.

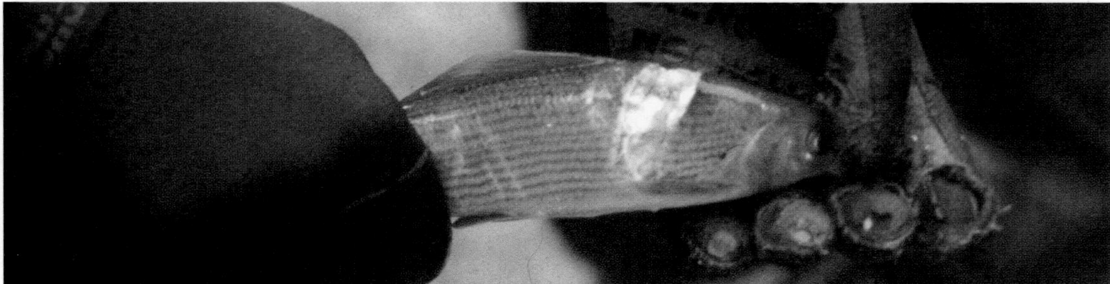

117 This wound was probably caused by a heron.

Sometimes, fish are found with wounds caused by a predator; heron stabs are not uncommon (117), characterised by a puncture wound on the back and, perhaps, elongated scars down the flanks of the fish. Whilst such a wound may eventually kill the fish, individuals are often seen in the river or are caught with apparently fully-healed scars. As an aside, trout taken on the rod bearing bird wounds will still probably be edible once the damaged area around the wound has been removed.

Fortunately, serious disease outbreaks are relatively uncommon. Unfortunately, except where a problem is identified early on and in an enclosed holding unit, effective remedial action is practically impossible for the most common problems found in chalk stream fishes. For example, the winter/early spring period is often marked by the discovery of trout and salmon bearing white or cream coloured patches of fungus, sometimes extensively covering the body. The problem is most apparent, but not exclusively so, in the cock fish, is most frequently related to changes in the fish's skin during sexual maturation and is caused by the fungus *Saprolegnia.* Lightly affected fish in the river may well recover of their own accord; there are specialist treatments for seriously infected fish in a fish farm situation though their efficacy is variable. If they can be caught, seriously affected fish in the river should be killed and disposed of appropriately.

An infamous disease which secondarily involves the fungus *Saprolegnia* is salmon disease or Ulcerative Dermal Necrosis (UDN). This is a spasmodic problem particularly amongst salmon moving into freshwater in the Spring and Autumn, but is also found in sea trout and brown trout. Typically, grey lesions appear on the head which may spread to form a complete 'cap'. Eventually, large ulcers may form and become reddish wounds which are subsequently invaded by bacteria and fungi. The fungus can spread to other parts of the body and lead to the fish's death. It is difficult to differentiate between a fish carrying UDN and one with a simple fungal infection; thus, one often hears fishermen and others describing fish in the river with UDN which may, in fact, be infected only with fungus. The cause of UDN is not known and, for fish in the river, no treatment is possible.

Amongst the most obvious and frequently encountered parasites seen on chalk stream fishes are the fish leech *(Piscicola geometra)* and the fish louse *(Argulus spp.)*. The fish leech is a red or brown worm-shaped organism, 1-3mm wide and 15-30mm long, with an obvious circular sucker at each end. The leech is a non-selective parasite, hanging on to submerged weeds awaiting the passing of a suitable host which may be as small as a stickleback or as large as a pike. Leeches tend to adhere to the underside of the host fish, but may be found on soft tissues such as the membranes of the eye or mouth.

The fish louse is a saucer-shaped crustacean, 2-5mm in diameter and largely colourless. Like the leech, the louse is not selective in choosing a suitable fish host, adhering to any part of the body, but particularly the fins.

Both the fish leech and the fish louse feed on the skin cells of the host. Normal, low levels of infestation of these parasites cause a fish few problems, but in large numbers they may create areas of bleeding under the skin and open the host to bacterial infection. Such secondary infection may, in extreme cases, lead to the death of the fish. However, as was emphasised earlier, a healthy fish in a river should not be sufficiently stressed, and thus vulnerable, so as to pick up excessive numbers of fish lice or fish leeches, and will live normally with a few of these parasites (say 1-10 in number). Therefore, the fisherman or fishery manager encountering a fish with a low parasite load should probably not interfere and attempt to remove the fish leeches or fish lice. Where a fish is caught which has a heavy parasite burden, it may be marginally the better option to remove the lice or leeches before the fish is returned to the water.

The control of predators and competitors

Most fisheries on the southern English chalk streams concentrate on maintaining conditions for trout (and salmon to a lesser extent). Therefore, any potential predator or competitor of the trout and salmon may be seen as undesirable and in need of control; it is important to appreciate, however, that *eradication* of the predator or competitor may well be practically impossible, so creating conditions which favour our preferred species might be the way ahead. The section below considers a number of predators and competitors of trout and, where applicable, possible control measures.

Mammals

Otter

As a fish-eating mammal, the otter is a possible predator of trout or salmon. On the Itchen otters have been reintroduced and may have spread to the Test, although they are seldom observed on either river. They will preferentially feed on those fish which are easiest to catch and, as such, the fast-swimming trout may be low down on the menu. Analysis of spraint of the otters introduced to the Itchen suggest that eels, bullheads, minnows, grayling and perch are prominent in the diet. Otters have been implicated in predation and damage of stocks of carp in fish farms, sometimes destroying fish worth individually several thousands of pounds. On these big fish, the square bite marks and scratches to the body can be very characteristic (118). However, this beautiful animal is very rightly afforded full and extensive protection under Euro-pean Directive and thus British legislation but control of the problem is possible with electrified, exclusion fencing around ponds, lakes and fish farms – this has been shown to be very effective (albeit expensive) where a complete fence can be built and maintained – otters have proved themselves very adept at finding even small gaps!

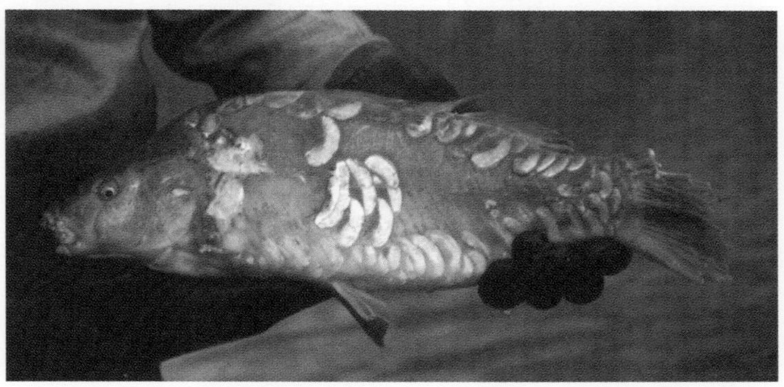

118 Bite marks and scratches on a carp, characteristic of an otter attack.

Mink

The mink is a North American immigrant which has established successfully in Britain. Mink take a wide range of prey including fish, birds and mammals. They are rarely seen, and a decrease in the local populations of waterfowl or small mammals such as voles may be the first indication of their presence. Mating takes place in late February-March; the young are born in late April or May, and stay with the mother until late summer before dispersing, travelling up to 48 km during the autumn and winter before settling in a new location. Individuals establish ranges, males occupying up to 2.5 km of waterway.

Mink may give clues to their presence in foot prints or in leaving dead fish on the river bank, killed by a bite to the backbone between head and dorsal fin. Droppings may be a tar-like sludge or be a firm sausage-shape, twisted and pointed, 60-90 mm long and 9 mm in diameter containing fur, feathers and fish scales. In warm weather, the dens (in sites such as holes in or beneath trees) may be detectable by their strong odour of faeces, decomposing animal remains and the mink's own scent.

Defra advise that, because of the damage they can do to wildlife, game, fisheries and domestic birds, mink are controlled when encountered. The most effective method of control is trapping, particularly during the mating season (February-March) and when the young are dispersing (August-October). Most commonly, cage traps are used, made of 14-gauge weldmesh and measuring 600 x 180 x 180mm. Some keepers bait the traps with fish, rabbit, pigeon, offal or tinned cat food, whilst others set unbaited traps positioned so that the mink's curiosity leads it to enter a dark, apparently interesting tunnel. Correct siting is vital: areas close to the water are favoured, perhaps under a bridge or where natural holes and tunnels will be investigated by the mink. Recent work by the Game Conservancy Trust has shown the benefit of mink 'rafts' which have a tell-tale pad of clay to show-up the tracks and thus presence of inquisitive mink. Once their presence has been confirmed, trapping efforts can be concentrated in those areas and possibly many mink caught at the same place over a long period. Traps should be camouflaged to avoid interference from the public and should be visited at least once per day. If a mink has been caught, it should be destroyed humanely, preferably shot by a .22 airgun or rifle or .410 shotgun whilst still in the trap. Drowning is not regarded as a humane method of disposal.

119 Game Conservancy Trust mink raft

Other mammals

Rats present problems to the fishery manager not because of any serious predation of fish (although they are known to bite fish in confined raceways), but because they act as a vector for the organism that causes Weil's Disease in humans and they create extensive burrow systems in the bank, aiding erosion of the stream margins. Where it is considered appropriate, attempts at control may be made by shooting, trapping or hunting with ferrets and dogs. Poison should not be used on the river bank.

Birds

The two most notorious avian predators of fish are the heron and cormorant. Whilst the cormorant feeds exclusively on fish, the heron will additionally take amphibians, small mammals and crayfish. Both birds favour fish 15-30 cm in length but will attack and injure significantly larger specimens, leaving nasty wounds. Cormorants, a major concern for fishery interests for over a decade, consume around 500 gm per day of fish and research by MAFF in the 1990's suggested that they can be extremely detrimental to stocks of riverine salmon and trout. Even owners and keepers on the upper Itchen, a shallow, fast flowing stream which should not be good fishing territory for cormorants report problems from the birds. Work by the NRA in the 1990's showed apparently significant predation on juvenile salmon by cormorants on the River Test, with, in one case, over 100 microtags retrieved from below one roost following that season's parr stocking.

Both the heron and cormorant are heavily protected under European and national legislation. As a consequence, it is illegal to shoot either species although licences may be granted by Defra's Wildlife Licensing Unit for limited culling (see www.defra.gov.uk). Therefore, any non-lethal control must be directed towards discouraging large numbers of either herons or cormorants. On the river, human presence is the only feasible form of deterrent.

Avian predation may be particularly problematical for those fisheries which hold or rear their own stock in ponds or raceways. Here, dawn and dusk raids by, for example, a heron or two may result in significant losses of fish. In these instances, nets or wire mesh screens may be used to cover the holding units. Alternatively, wires 30-60cm apart may be strung above head-height over the holding units although herons will still drop through such defences.

A large number of other bird species may predate on trout and salmon fry (e.g. kingfisher, dabchick, coots) or compete with the fish by feeding on similar invertebrate food items (e.g. ducks, moorhens). However, such species have an equal part in a balanced ecosystem and probably cause insignificant problems in a fishery. Some other species, such as the Canada goose and the swan, create disturbance to the angler by their physical size and mess on the bank through their not-inconsiderable faeces. Swans are also responsible for what many believe to be a very major threat to the ecology of the southern chalk streams in their large-scale grazing of extensive areas of aquatic weeds in many southern English rivers, including the Test and Itchen. The lower Test, for example, has swan flocks numbering more than fifty birds. Again, however, the law favours the bird and no more than discouragement is the order of the day.

Fish

Pike

As outlined earlier, the pike preferentially preys upon fish 10-25% of its own weight and consumes around 3% of its bodyweight per day. The ratio between the weight of pike and prey fish in a water is normally in the range of 1:4 – 1:8. They are non-selective in terms of the fish species taken and therefore the trout features in the diet of chalk stream pike. Probably, some control of pike numbers makes sense where stocking is the norm because stocked fish of (say) 600gms fall well within the preferred prey size of pike larger than around 2kg and methods like electric fishing (see below) or angling can be effective in keeping on top of pike of this size. Pike control is somewhat more controversial in wild trout fisheries, with the 'cull' and 'no-cull' camps strongly divided. There is evidence to suggest that wild trout numbers can actually be depressed by attempts to control the pike population because methods used to remove the pike (electric fishing or angling) tend to be selective towards larger fish. These are the very fish that may well regulate, through cannibalism, numbers of their smaller brethren. So, in the absence of the larger pike removed by the fishery manager, survival rates of small pike increase and these fish prey more heavily on young trout – thus, overall, trout numbers go down. In shallow, narrow river sections, where electric fishing crews can fish well and catch pike of all sizes (even the young-of-the-year fish, around 10 cm in November) some control of pike numbers might be a good idea though plenty of eminent fishery people suggest that pike should be left alone to self-regulate.

Eel

Together with some other river fish like the perch or chub, the eel occupies a position as a potential competitor and predator of trout. Eels are said by some to be particularly predatory on salmonid eggs and alevins, but since the eel feeds very little in winter (because of the low water temperatures), this seems unlikely. In most studies, winter-caught eels have been found to have empty stomachs, although one study in Lake Windermere did find Arctic Char eggs (spawned from October-March) in the diet. The current crisis in eel numbers makes this fish even less of a threat for the sporting salmonids of the chalk streams.

Grayling

There remains considerable debate about how closely grayling and trout share common river space and food items; there is evidence of overlap but also evidence of subtle segregation of both diet and habitat. The grayling is a fecund species, capable of a rapid increase in numbers, though also, paradoxically, of rapid *decline* in numbers in some fisheries. It may be too that very large grayling become predatory, taking eggs and juveniles of other species; in Europe, big grayling are caught by anglers using small Mepps-type spinners. The grayling is a very free-rising fish and therefore if a fishery is aiming to provide trout fishing, excessive numbers of grayling may become a nuisance to the anglers. For these reasons, autumnal removal of huge numbers of grayling was commonly practised; thankfully nowadays, however, many fisheries exploit grayling fishing in the winter and grayling electric fishing days are not what they were twenty years ago and more.

Methods of control of unwanted fish species

Angling

The use of conventional rod and line to remove unwanted fish species offers a number of advantages. For example, it is possible to be extremely selective in the removal of small numbers of unwanted fish using a rod – a few pike may be taken on a spinner in a short time by one man, avoiding the paraphernalia and potential disturbance to the fishery associated with an electric fishing operation. Indeed, many chalk stream fisheries offer winter coarse fishing as a bonus for the club members or on a day-ticket basis. Where wild trout are present in numbers, there is a case for either finishing coarse fishing by about mid-November or not allowing it until late January to avoid possible damage to any accidentally caught trout which are preparing for or recovering from, spawning. Even so, it is prudent to impose a sensible set of rules on the coarse fishermen by, for example, insisting on the use of barbless hooks. In this way, trout (or salmon) which will inevitably be caught from time to time may be returned unharmed. The potential benefit to the fishery of a good head of coarse species may be realised when the keeper can barter favours in exchange for a day's fishing of quality chub or roach!

Netting

Nowadays, netting techniques are not used on the chalk streams for removal of unwanted species, although they were used to harvest grayling on shallow, fast-flowing reaches of the upper Test as recently as 25 years ago. Electric fishing is now the preferred method for those fisheries which still carry out winter removal of pike, grayling and maybe left-over stocked trout. The use of any fish removal equipment must be licensed or consented to by the Environment Agency prior to use.

Fish traps

On the chalk streams, the use of traps is confined to the capture of pike and eels (although in his book Fishery *Management and Keepering,* the late Richard Seymour shows a good trap design for collecting trout broodstock). The most simple designs, such as baited pots or fyke nets, operate by directing the fish into the trap through a confined entrance. Whilst basically simple to use, such traps require expertise to be used effectively and are reliant on the fish being active. In flowing water, they will collect debris and may subsequently be washed away. Furthermore, traps may also take creatures other than the intended quarry, including salmon and trout.

The most sophisticated types of trap include the eel rack or Wolff grid which effectively 'strains' downstream-migrating fish from the river. The entrance to the eel rack is ordinarily closed by a set of upstream hatches. When the trap is fished (by night), the hatches are lifted allowing the river water to flow over a grid which filters out all flotsam and jetsam, including fish. Weed and debris is raked off the grid whilst the catch of eels is gathered up or channelled into a collecting box. On really good nights, these traps can take hundreds of pounds weight of eels, though, at least for now, those days are gone.

Under the terms of the Salmon and Freshwater Fisheries Act 1975, the use of any trap, whether a small crayfish pot or large eel rack, is illegal without a licence from the Environment Agency (see www.efishbusiness.co.uk for further advice).

Other methods

A variety of other methods has historically been employed, principally against the pike, including shooting, harpooning and snaring or 'wiring', in which a running wire noose is mounted on the end of a 4-5m long bamboo pole and slipped over a basking pike; a firm jerk tightens the noose around the pike and the fish is dragged out. The Salmon and Freshwater Fisheries Act 1975 now forbids the use of these methods.

Electric fishing

Electric fishing is regarded by many as an essential fishery management tool not only in the live capture of unwanted species but also in stock assessment and broodstock capture; to others, electric fishing is destructive of trout, salmon and their invertebrate food items. The scientific literature is equivocal, though much recent work highlights the potentially injurious nature of electric fishing to the individual fish (rather than at a population level), even if sometimes on the back of truly poor science! From a personal stance, based on reading the science and 25 years of practical electric fishing experience, I believe that the method has a place where a specific job needs doing and where no other method will work successfully *and* where simple precautions (discussed below) are taken to optimise fish welfare.

Electric fishing is not new – the first patent for an electric fishing machine was granted in 1863, but major advances in the technique have been made only since the Second World War and especially in the last decade.

The basic principles of electric fishing, in very simplified terms, are that a power source (i.e. a generator) produces electricity which is regulated by a control box before being passed into the water by metal electrodes; fish moving into this electrified zone are affected, possibly being drawn by an induced swimming action before becoming stunned and hopefully caught in a hand net. The area of water which immediately surrounds the electrodes contains the strongest electrical forces which dissipate with distance from the electrode. Furthermore, the amount of electricity that flows through water (and the fish) is dictated by the conductivity of the water which in turn is dictated by the quantity and type of materials dissolved and suspended in it. Distilled water will, theoretically, have a conductivity of zero whilst sea water is highly conductive; chalk streams are normally high-conductivity waters and therefore lend themselves well to electric fishing.

Effects of electricity on fish

Electric fishing produces a range of responses in a fish by working on its nervous system or muscles: escape if the electrical field is weak, involuntary swimming towards the electrode if the voltage is high enough or muscle paralysis close in by the electrodes. The amount of electricity to which a fish is exposed decreases with increasing distance from the electrodes and thus, at some point, the fish is perhaps aware of the electricity but is not positively influenced by it. This suggests that electric fishing has a limited range, generally not greater than around 2m, usually much less.

Large fish are more likely to succumb to electric fishing through a combination of some greater effect from the electricity and size-selectivity by the operators. Some species are widely thought to be easier to catch (pike being one of them, luckily for some river keepers) and some species are certainly more vulnerable to harm from electric fishing – trout, salmon, grayling, chub and dace are high up in this list.

Electric fishing can cause harm to fish in a number of ways. Most dramatically, fish can suffer dislocations of the spine from severe, simultaneous contraction of the muscles of the body. This in itself may not be fatal, but if the dislocation ruptures the major blood vessel running by the spine of the fish, it is very likely to die from internal haemorrhage. One very eminent scientist says that spinal dislocations are always accompanied by bruising on the skin above the point of the dislocation – these bruises are what are commonly called 'burns'. More subtly, electric fishing can cause nervous fatigue which means that vital systems in the body fail such as heart or gill function.

Sadly, these negative aspects of electric fishing have to be accepted as a possibility and the method only applied if the job at hand is deemed sufficiently important. A number of really useful, practical steps can be taken to optimise the welfare of captured fish. Power output from the equipment should be set *just* high enough to catch fish – cranked-up kit at high voltages and amperages kills fish and very often is not needed. Modern equipment allows operators to fish with pulsed direct current (PDC) electricity set at low frequencies – this is known to be less harmful to fish. Smooth or straight direct current (DC) is even better but can make fish catching difficult. Alternating current (AC), still used by some, is definitely very damaging for fish. Operators should get used to turning off electrodes when fish come close and are obviously stunned, just long enough to let netsmen scoop up the catch which should then be turned out quickly into *oxygenated* recovery tanks. In turn, recovery tanks should be emptied when no more than half full, not when it is no longer possible to fit in another, single fish! This regular emptying regime also encourages regular water changes, another good practice.

Effects of electricity on invertebrates

The effects of electricity on invertebrates is poorly understood. Work 30 years ago on the Test and Itchen suggested that the effects were largely confined to short distance, downstream displacement and recent work in Scotland has shown no short term physiological effects in one type of invertebrate, the pearl mussel. However, electric fishing has been used scientifically to catch crayfish, so it does affect at least this group of large invertebrates.

Until the potential effects on invertebrates becomes better understood, these should again be considered in the assessment of the need to carry out any electric fishing job.

Field use

Preparatory work for an electric fishing session should include gaining permission from the Environment Agency (see www.efishbusiness.co.uk for consent forms), reconnaissance of the site for access and depth, a thorough pre-start check of the equipment, a briefing of personnel and an extensive risk assessment of the job.

The Environment Agency has a Code of Practice for Electric Fishing which, whilst written specifically for its employees, could be regarded as the industry standard in terms of application and health and safety. Certainly, the Code includes much of relevance to any electric fishing team. For example, it stipulates the need for equipment checks, safe working systems, sensible situations for wading or boat work and procedures in the event of accidents. Some of the detail may differ for the club member or keeper but the Code's principles are sound.

Wherever possible (in water on average at or below thigh depth), electric fishing is most efficiently done by operators wading in the water in chest waders (with life jackets) or dry suits. These crews work upstream, sending muddy water behind them and using the flow to bring stunned fish back to the netsmen. In deeper water, the crew should take to a boat which can either be motored upstream or drifted downstream, roped from each bank. Some keepers favour downstream boat work, especially for grayling removal, because, they say, catch rates are higher – my experience *might* support this!

In conclusion, whilst electric fishing does have a number of drawbacks in certain conditions, and it is not favoured by some owners and keepers, it can have a place in the management of many chalk stream fisheries.

120 Martin Murrell, Chairman Hampshire River Keepers, with the tool kit for today's river keeper
Photo: Ian Wrightson

Chapter Six

THE RIVER KEEPER'S YEAR

The river keeper's year is inextricably linked to the ever changing seasons and the changing moods of the river. Working with nature's rhythms and understanding the needs of the river, its floodplain and the wildlife that inhabit our chalk streams, is central to the life of a river keeper. Such expertise and experience is gained through many years of living and working on our chalk streams. For some keepers it is in their blood with knowledge and tradition being passed down from one generation to the next, in families such as the Lunns, Hills and Walfords. Today there are over sixty river keepers on the Test and Itchen who care for these famous rivers making each year's fishing season an exciting and memorable experience.

It could be said that when the fishing season ends the river keeper's year begins. The success or failure of the next season for many anglers and owners will depend largely on the endeavours of the keepers during the winter months. To the untrained eye much of the work undertaken by river keepers can go unnoticed. River bank repairs that can take weeks to complete will be covered with a fresh growth of herbage by mayfly time. Banks of mud and silt brought down river by the autumn rains will have been cleared from spawning sites and bank side willows cut back in the winter will be covered with fresh shoots. These seasonal tasks are reflected in the diary of a typical river keeper.

Winter work

The legal end of the trout fishing season on the Test and Itchen is the 31st October, although many fisheries close at the end of September or the middle of October. However coarse and grayling fishing, which is becoming increasingly popular, can continue until 15th March. Visiting anglers often ask what river keepers do during the winter months. They are busy looking after the river in preparation for the coming fishing season and many keepers are also involved with the shooting season.

Leaving the fringe

Traditionally the river's fringe has been cut to 'put the river to bed', however there is an increasing recognition that if left uncut it provides invaluable cover and a refuge for fish from predators during the winter months after the river weed has died back. The fringe is also a vital habitat for a range of species that enrich the chalk stream fishing experience.

Managing grayling and pike

Historically species such as grayling were removed by electric fishing, however today it is recognised a better balance of fewer but larger fish results if the stocks are left to settle down. With cormorants also taking a toll and a requirement to maintain adequate stocks for winter grayling fishing, large-scale culls are no longer undertaken. It is also recognised that removing large pike can lead to a proliferation of smaller pike and it is sometimes better to leave nature to establish the right balance of fish stocks.

Autumn weed cut

The autumn weed cut is carried out to help the river to scour, particularly in times of low flows. It also reduces the likelihood that root stocks will be washed out during times of flood.

Cleaning spawning gravels

Many owners, together with the Environment Agency, clean and loosen the gravel to promote successful salmon and trout spawning. Research projects have shown that egg survival can otherwise be very low due to the suffocating effect of fine silt smothering the eggs. Much improved fry emergence can be obtained in stretches where the gravel has been cleaned by raking or water jetting.

121 Gravel cleaning *Photo: Mark Sidebottom*

Ditch clearance and tree management

The next task is clearing ditches and tree management, which includes coppicing and pollarding bank side trees. This is particularly important as trees which are not managed in this way have a tendency to fall in high winds, often creating a large hole in the river bank.

Vermin

The siting and daily checking of traps and cages to keep vermin under control can take up a good deal of time. For mink control, the Game Conservancy has developed a raft with a tray of clay which is kept permanently moist. Any visiting mink can then be detected by their footprints and a trap substituted, which must then be inspected on a daily basis. Once a mink is caught the clay tray can be put back to see whether any more mink are in the vicinity and inspection can be reduced to once a week. Using this method mink have been virtually eliminated from the Itchen. It is a credit to the river keepers, professional and part-time, that so few mink and rats are noticed by anglers.

Cormorant predation particularly on juvenile salmon and brown trout is a cause of concern to everybody involved with fishery management. Licences to shoot cormorants can be obtained from Defra to prevent serious damage to fisheries. Licences will only usually be granted when non-lethal anti-predation measures have been tried. An official from Defra will visit a fishery to verify the information which has been given. The cormorant problem usually starts in October and ends before May. If an SSSI lies within your boundaries or adjoining lands are covered by such designations Natural England must be notified.

River bank maintenance

Recent drought years such as 1989-2, 2005-6 and the flood conditions of 1994-5, 2000-1, have demonstrated the need to manage the rivers for both extremes. They have also shown the vital importance of maintaining hatches, keeping flood ditches open, and defending river banks, but also of providing much needed habitat and cover for times of low flow.

On chalk streams that have been modified by man over the centuries, areas particularly vulnerable to erosion are pressure points immediately upstream of weirs and hatches and the turbulent downstream areas. Cattle can also contribute significantly to bank erosion and ruin a river bank. Fences should be checked prior to turning them out.

Planning – Need to obtain consent

The Environment Agency requires owners to seek a consent for any works in or within 8 metres of a main river, so that flood risk is not exacerbated and important habitat is not damaged. If any of the proposed work lies within an SSSI; consent from Natural England will also have to be obtained in advance of starting the work. Some minor repairs to river banks may be undertaken without the need of a consent, however works such as reveting the bank, narrowing the river or digging new pools would need Environment Agency consent. Planning permission may also be required for any major work. If owners are unsure of the position, a telephone call to the Environment Agency is advisable.

Timing

The most suitable month for river bank repairs is October when the chalk streams are usually at their lowest. Major works should be planned so that they do not coincide with salmon and trout spawning, which takes place in December and January.

Traditional river bank repairs

Broadly speaking the shallower the water under a river bank the cheaper the cost of any repairs. Other factors such as access and availability of suitable materials are also vital.

Let's assume a section of bank has eroded and needs attention. The water under the bank is two feet deep. The first task is to find out if the bed of the river is firm or soft. Don't be deceived by appearances – always check. Often gravel can be thin with a layer of under-lying peat below. River keepers keep an iron bar around six feet long with a diameter of two inches for such tasks. Once the depth is established, posts of suitable length can be brought to the site. A post is driven in at each end of the repair, and a string line is pulled between them. Further posts are then driven in along the line at three foot centres. If the gravel is compacted, a hole is made by knocking in the iron bar and putting in the post down the pilot hole. Posts should always be hammered in with a post driver (bumper), never a sledge hammer as the posts may split. Willows or hazels that have been pollarded or coppiced in the past, would then be cut down again for their straight shoots. All side growth is trimmed away and saved. Bundles of shoots ten to fourteen feet long with a diameter of one and a half to two inches are taken to the repair. They are then wound in and out of the posts and pushed down to the river bed. This can be done quickly and when the top of the posts are reached a hurdle effect is achieved (122). At this stage if the river bed was soft, a tie of straining wire should be made to every third post. A further post is then driven in the bank behind the straining wire post, and the wire is tensioned. Remember to keep all wire ties about a third down from the top of the post. All side-growth and other suitable branches etc. that were saved are cut into lengths of four, five and six feet and tied into bundles known as faggots with binder twine. The various length faggots are then placed behind the repair and packed in tight. If green willows are to be used, beware, they can root. To combat this tendency, make them up in advance of any bank repair.

122 Posts and willow spiling are used to repair an eroded bank *Photo: Guy Robinson*

A layer of cut reeds or sedge is then spread on top, followed by loads of suitable backfill material (123, 124, 125). Chalk, if available, will generally provide a longer-

123 Hazel faggots are used to create the outline of an island *Photo: Richard Redsull*

124 Chalk back-fill consolidates the island *Photo: Richard Redsull*

125 Top soil is spread over the chalk to encourage marginal plants to establish *Photo: Richard Redsull*

lasting repair than soil which tends to erode again behind the hurdle. Finally a thin layer of local soil is spread over and raked in.

Hazel faggots can also be used to front bank repairs. The faggots will last about 4 years and provide habitat for aquatic invertebrates. The use of coppiced hazel also supports a traditional rural activity and creates the ideal habitat for bluebells.

To give nature a helping hand nearby marginal river plants can be planted to accelerate bank re-growth and tie in the backfill material. Freshly cut willow can also be successfully planted in the winter to help consolidate the bank and provide cover for fish. Cover can be further provided by submerging a branch of a tree alongside the bank repair which will provide a refuge for all manner of shrimps, crayfish, eels and, of course trout. A useful reference for river habitat enhancement work is the Wild Trout Trust's, *Wild Trout Survival Guide (2006)*.

Mudding

Mudding is a traditional technique used by some river keepers but is considered by many to be potentially damaging and unsustainable. Where mudding is considered necessary it is often an indication that the channel is over-wide or deep for the given flow. Consolidating areas of deposited silt with the use of hurdles or groynes and then planting with emergent plants is now viewed as the best option. To reduce the levels of silt on the river, once the fishing season is over, hatches that have been lowered to maintain water levels should be raised wherever practicable to encourage the river to self-scour its channel. Where possible close liaison with the nearby farmers can help ensure that silt does not run-off the surrounding fields. A buffer strip along the river's margins is an invaluable protection against run-off and will also help absorb any fertilizer or pesticides in the run-off.

Mud and silt deposits often accumulate very quickly during the winter months. The crystal clear waters of the chalk streams during the summer months can turn to a colour reminiscent of cocoa during the winter. The mud problem is far more difficult to control in the lower Test, Itchen and Avon valleys than it is in the faster flowing upper regions. Large amounts of mud will settle out quickly in waters with little flow. Such silt deposits are a natural part of a river's process of erosion and deposition, however there are concerns that with the intensification of agriculture it is worse than it used to be. Where the river is wadeable many keepers with inner bends which regularly get banks of mud deposited under them, strategically place sheets of corrugated iron to deflect the current into the area which is known to silt up. This is done in advance of the winter rains, using fence posts to brace the structure. The sheet and posts should be removed before the fishing season opens in the spring.

The mud can be removed using mud-pans formed from a steel plate resembling a reversed dustpan attached to a long handle and deposited onto the bank. Holes should be drilled into the base of the pan to allow water to drain out.

Major dredging and mudding operations should be completed by Christmas so as not to damage trout and salmon eggs incubating in the river's gravel.

Punts

Bank repairs occur in the most awkward places. Often the only access is by foot or by using a punt. Punts can be used for heavy work such as carting loads of chalk down river, but they need to be of a rugged construction. Fibre glass fishing punts are not suitable. A sheet of exterior plywood should be laid on the floor of punts to protect them when doing work of this kind, it also makes shovelling the chalk out easier. Punts often have rings fitted to the front and rear. When pulling them upstream, two people are needed, one to pull and one to fend the punt around bank-side obstacles. A fifteen foot punt should also have a further ring attached to its side approximately five feet from the front. When a rope is fixed to the ring and pulled the punt will go out into the river on its own, and this can be useful when working alone.

Bridges

126 Footbridge over the River Test *Photo: Lawrence Talks*

When budgeting for a winter work programme, repairs or replacements of foot bridges are often a major item (126). Before renewing a bridge over any river, thought should be given about its intended use in the future. The design and proportions of bridges suitable for anglers would certainly need up-rating before machinery such as mowers or mini tractors for example could safely use the bridge. The finished height at the underside of the bridge should be well above the known high water mark. Other considerations would be whether there will be a need to get a weed cutting boat underneath the bridge, or an electro-fishing team etc. It may be a good idea to incorporate a lift up section in the bridge. Always allow enough length for the bridge to continue a little over the bank at each end to combat any bank erosion. Often the bridge that is going to be replaced can be used as a platform, and the new bridge erected around it. Piles for bridges are usually oak. Remember when ordering to state that you want them pointed. It is important visually that bridges are put up at right angles to the river. Before ordering any timber check the river bed with an iron bar to establish the lengths of piles required. Oak is the most suitable timber for bridge building because of its durability and great strength. Costs can be reduced by using larch or Douglas Fir for the planks and hand-rail.

Apart from minor repairs, most bridge replacement work will require consent from the Environment Agency.

Fences

Where cattle or horses graze in river meadows, fences should be erected to protect the banks. Ideally the fence should be at least the width of the river away from its bank-line, but often a compromise is reached between fishery and agricultural needs. A good fence should comprise at least three and preferably four strands of wire, with straining posts at intervals of fifty yards and intermediate posts every ten feet. The whole construction should be tensioned and properly stapled. Remember to leave access for mowers and other machinery. It may also be desirable to install stiles at intervals along the fence line in order that anglers are able to walk back down stream, well away from the river bank. Cattle drinks should be sited in places where the ground is composed of gravel and is stable. The fence line should be continued into the river to prevent animals straying. Pasture pumps can be installed where access to the water for livestock is undesirable. Portable electric fences can also be invaluable in managing limited grazing to promote suitable meadow plant species while at the same time protecting soft margins.

The fishing hut

The fishing hut is an important feature of any fishery (127). Some large estates have a hut on each beat, others prefer a large communal hut which can promote a friendly and jovial atmosphere. If possible the hut should be sited with a view of the river and be within a reasonable distance of the car parking area. Unfortunately some commercially made garden sheds and chalets are expensive despite being of a flimsy construction. Many owners and keepers prefer to make their own, usually buying in sawn timber direct from a local sawmill. Planning restrictions apart, every effort should be made to ensure that the fishing hut blends in and enhances the fishery. Try to visualize the finished hut on your chosen site before building commences. In days gone by, fishing huts were built on top of a few barrow loads of chalk. A bench seat is usually fitted in the hut and often folding chairs for use outside are a good idea. Many fisheries keep a return-book in the fishing hut, some also have an observations book and notice board.

In some locations the security of the hut and its contents will require serious consideration. The door should be furnished with a good quality mortice lock or even a stout padlock. All securing screws or better still bolts should be fitted so as to be impossible to unscrew from the outside. The best protection for windows are shutters. These can be fitted so as to either hinge up and down, or to slide up and down in runners, both arrangements being secured inside by a large bolt which passes through the closed shutter into the interior of the hut. The shutters themselves are best made from very strong wood or even steel and obviously fitted to every window. Where these precautions are necessary, a daily routine for opening and closing-up the hut will have to be established. Even with the most stringent precautions, vandalism and break-ins are unfortunately a feature of today. It is recommended that the minimum of equipment is kept in huts and no inflammable materials.

Unless an existing fishing hut is being replaced, planning permission may be needed and consent is also required from the Environment Agency if the hut is within 8 metres of the river bank.

127 Guy Robinson, Leckford Head River Keeper outside thatched fishing hut on the River Test
Photo: Lawrence Talks

Hatches

Hatches and water level control structures exacerbate siltation problems and delay fish on spawning migrations. Classic chalk stream habitats always develop more quickly on sections of river which are left to flow as fast and shallow as possible. There should always be a presumption to leave hatches as fully drawn as practically possible. In-channel and marginal habitats will always respond better in sections where there is the minimum interference with water levels (128).

When the flow in chalk streams is very low, it is possible for hatches to be lowered to maintain upstream river levels. It is important, particularly during the autumn, on large weirs and major hatches, to ensure that migratory fish have free passage. One gate raised fully offers salmon and sea trout a far better chance to ascend unharmed, than a number of gates opened a few inches across the width of the river.

Hatches not only control the water velocity, but water levels as well. On side-streams and small water courses this is usually achieved by using sawn oak hatch boards of various widths, which are often held in position by water pressure. The construction and repair of major hatches across a river is a highly specialised area and professional advice should be sought before attempting to repair or build such structures. Any obstruction to river flows or the diversion of a water course would need Environment Agency consent. Established small side-stream hatches should be renewed if they are falling into disrepair.

It is sometimes easier to construct a replacement hatch a few yards up or down stream of the existing hatch. On completion the disused hatch can be pulled out and the area reinstated. The hatch will last many years if built on a concrete base.

128 Adjustable hatch *Photo: Lawrence Talks*

Stiles and gates

Whenever a fence or hedge needs to be crossed a stile or gate is the usual means of access. Always look for ways to keep the number of these to the minimum. Many fisheries have members whose increasing years make clambering over stiles very tiring and sometimes dangerous. A better option is the walk-through stile, with the kissing gate being a more expensive version. Where access is also required for vehicles and machinery mowers etc., a gate must be provided. Slip rails provide a cheaper and satisfactory alternative to the conventional five-bar gate. Short detachable lengths of fencing wire slung between two posts provide temporary solutions, that invariably become permanent – the traditional Hampshire gate.

Fishing seats

Fishing seats (129) apart from typifying the chalk stream fisheries are a positive aid to correct chalk stream fly fishing. If carefully located, they will encourage the fly fisherman to progress slowly upstream resting upon the seats in order to adopt a low profile and view the length ahead of him. This leads to fewer disturbed fish and enhances the possibility of trout actually coming on the fin. Seats should be sited in positions which give the angler a good view of the water. A seat erected at the bottom of each beat looking upstream provides anglers with a suitable meeting place.

129 Leckford fishing seat complete with notch to rest your rod *Photo: Lawrence Talks*

Summer Work

Fish stocking

In spring, in preparation for the new fishing season, trout are stocked into the river if stocks of wild fish are insufficient to support the fishery. The river is now emerging from winter, with beds of emerald green *Ranunculus* beginning to appear and spring flowers heralding a new fishing season.

Orders have to be placed early to source brown trout that have been reared in the Test and Itchen as they are often in short supply. It is usual to give suppliers of trout an estimate of numbers required and size of fish well before Christmas, for the oncoming season. Exact details can be firmed up when delivery dates are arranged in the spring. Stocking with all female triploid brown trout reduces the risks of stocked fish interacting with wild fish and will enable the wild component of the stock to develop. To reduce the potential of introducing any unhealthy fish into the river, it is essential that the fish have been health checked and it is a legal requirement to secure written Section 30 consent from the Environment Agency in advance of the trout being stocked. To keep disturbance of freshly stocked trout to a minimum any weed cutting should be completed before trout are introduced. Submerged wooden boxes (131) are often used to introduce a few fish every few yards through a fishery, to ensure a widespread distribution of the fish. Trout are naturally territorial and are more likely to settle if spread well out in lower densities. Once released the trout soon find a lie from where they sip spinners or rise for an emergent fly sailing past, trapped in the river's flowing waters.

130 Fish transporter *Photo: Lawrence Talks*

131 Fish boxes are used to stock trout along the river *Photo: Guy Robinson*

Weed cutting

Luxuriant beds of chalk stream plants or 'river weed' move with the river's flow in its crystal clear waters, captivating those who stop and stare.

Managing chalk river weed is an essential job of a river keeper who needs to ensure good fishing, cover for fish and a patchwork of habitats that support the river's varied fly life and species such as the water vole which feeds on the rich and varied larder of river plants that the river provides.

River weed is vital for the ecology and well-being of a river. On gravel shallows the weed provides much needed cover for juvenile wild trout and salmon from predators. In times of high flows a healthy weed growth on shallows helps to consolidate the river bed and prevent erosion. In the summer months, watching juvenile fish moving through weed beds gives great pleasure and is very rewarding.

Ideally every yard of river contained within a fishery should have a succession of weed beds which are healthy, vigorous and as far as possible comprise all the most beneficial plants. The river's weed diverts, slows and concentrates the river's flow. Where water is forced to flow around a bed of weed it creates localised areas of quicker water, which can create clear runs of gravel on riffles or cause the river's energy to scour the margins. Where a weed bed has been allowed to grow and develop, it will filter out finer materials until a large area of mud and silt is consolidated around the basal stems. Some keepers call these 'permanent' weed beds.

Great care is needed when weed cutting as it is very easy to cut too much weed. Weed cutting requires careful forethought with local knowledge of past weed cuts and the effect they had being an invaluable reference, which develops with time spent on the river, for no two years are ever quite the same for weed growth.

Before starting it is important to have a clear idea of which weeds are to be cut and which are to be left. In times of drought weed may need to be left to grow to help maintain a head of water. During prolonged periods of high water weed may need to be cut hard to reduce levels that if left unchecked could cause flooding of nearby riverside houses.

Most rivers have an agreed system of dates for weed cutting so that cutting is restricted to specific times and the nuisance of floating cut weed is kept to a minimum. Cutting of weed requires a consent from the Environment Agency under Section 90 of the Water Resources Act 1991. On the Test and Itchen an authorisation for cutting dates for the whole of each river is obtained by the Test and Itchen Association and communicated to all owners and keepers.

There's an old saying, "Look after the river under the banks, and the middle will look after itself". If beds of weeds are left in the centre of a river, the velocity of water under each bank can be influenced by the width of channel cut.

If the opposite river bank is not managed by you, a joint plan will have to be agreed well in advance of any weed cutting. This is vital as cutting weeds on one side of a river pulls the water over from the other side. Similarly your upstream neighbour's water levels could be dramatically reduced by your actions.

It is most important that an immediate start is made on day one of the cutting dates. This allows plenty of time for your cut weeds to be carried away downstream. Weeds that are cut on the last day of a cutting period can be a nuisance to fisheries below you. The last two days should be used to clear away hung up weeds etc. During heavy weed cuts, many professional keepers faced with acres of cutting, start work at first light. This gives them a good start before cut weeds start reaching them from upstream.

On the Test and Itchen the weed cutting periods typically last for seven to ten days and occur in April, June, July and August. Woe betide anyone cutting weed outside the pre-determined dates as this is viewed with great disdain as it interrupts fishing downstream. 132

Weed cutting is a strenuous activity and should only be undertaken by persons who are physically fit. In deep water life jackets are essential. Always tell someone the area in which you are going to be working, and if possible work in pairs. As you work, keep a watchful eye behind you, particularly when wading deeper waters, as large rafts of weeds can easily knock you over. If for any reason you start to feel unwell, get out of the water straight away. Always wash your hands after working in the river, because of the risk of Weil's disease and have a first aid kit close to hand.

132 Paul Moncaster clearing down after weed cutting on the Test at Tufton. *Photo: R H Dadswell*

River keepers and aquatic plants

Water-crowfoot *Ranunculus* is our most common and important river weed. It requires moderate to strong flows and favours a gravel bed. It grows until it flowers producing white buttercup-like flowers for the species have a common ancestry. It then starts to die back. If cut before flowering it is stimulated and starts to thicken up again. Sometimes a light topping is all that is necessary to keep it in check until the next cut.

Water-starwort can play an important role in filtering out mud and silt. It is often found in slow-flowing water where it provides important cover and habitat for invertebrates, particularly shrimps

Water-celery (also known as Lesser Water-parsnip) and Fool's Watercress are as beneficial to the chalk streams as Water-crowfoot. They are excellent oxygenators, and their dense growth harbours a myriad of invertebrates. In shallow waters, beds of Water-celery are sometimes left and weeds around them cut. This causes a bore effect as water is forced around and over that bed.

Mare's-tail tends to grow on mud and silt in slower flowing reaches. It can become quite dense. The emergent stems of the plant look like little Christmas trees. Pike will spawn in areas of Mare's-tail in January and February.

Pond weeds grow in moderate to slow flowing stretches of water. There are a number of species, some become dense and difficult to clear if left to emerge. One of the broad leaved variety can grow to over sixteen feet in length.

Watercress harbours an astonishing variety of nymphs, fish fry and snails providing much needed marginal cover. Uncontrolled cress beds can quickly grow from bank to bank on smaller streams or concentrate the stream's flow into the centre of the channel providing fast flowing water which harbours stonefly nymphs. The growth can also provide sheltered areas for mayfly larvae. In fishing areas, cress can be managed by pulling some of it onto the riverbank with a pair of grabs or a four grain prong at an early stage.

River Water-dropwort, also known locally as carrot-weed, is often found growing with Water-crowfoot. Its distribution is patchy, but it is generally agreed to be of great benefit to a river. It is easy to manage and holds many species of invertebrates. Often leaves, twigs and fibrous matter settle out behind this plant, making it a favourite haunt of mayfly nymphs. However it should be recognised that this plant when pulled from the river and allowed to dry out on the river bank is extremely *poisonous* to livestock.

Bur-reed, which is often called ribbon weed by Hampshire keepers. These plants are difficult weeds to cut and need careful control on trout water. A stretch of river with Common Club-rush present can quickly become unfishable, although on some slow flowing stretches it is the only weed present and provides welcome cover for fish.

Technique

A few days before a weed cut commences, river keepers have a good look at the weed growth, and plan their weed cutting programme. Usually main river first, followed by any carriers and finally side streams and ditches. The time factor is the only reason river keepers follow this sequence, for example the July and August cutting period are usually only a week each month. The most important and time consuming is the main channel, so this is completed first. Always note the water level before starting particularly if you are unfamiliar with the river. Perhaps a chalk mark could be made on a bridge, or a peg could be pushed in the bank at some prominent point. As you work keep an eye on the mark to ensure you are not reducing the water to an unfishable level.

Most river keepers today cut wadeable water with a Turk scythe (133). Wearing thigh or chest waders walk downstream cutting from right to left as you go. The point of the blade just skims across the river bed in front of you. For each step that is taken, a cut is made. Don't try to cut too much at first, just work without stretching or reaching out. The water will soon colour up as silt is released from the cut weeds. With practice it is not difficult to work blind as you will be able to feel the weeds with the scythe. A carborundum stone is used to sharpen the blade and frequent stops should be made to retain a good edge. On sunny days wear polarized sunglasses to reduce the glare of the sun as it reflects on the water.

Line and bar cutting – also called side and bar

The line and bar method of weed control has two roles, maintaining maximum water levels on one hand and efficient weed control on the other. It is only carried out on easily waded water, usually with a healthy growth of *Ranunculus* and Water-celery. If

133 Weed cutting with Turk scythes *Photo: Guy Robinson*

134 A pole scythe being used to cut weed in deep water *Photo: Guy Robinson*

starting from scratch, a careful assessment will be needed to evaluate the most strategic positions for each bar. Often the bars are encouraged in wider shallow areas, and weeds are cleared in the narrow deeper water where trout may hold better. To form the bars, weeds are allowed to flourish from bank to bank. The length of each bar can vary but should be at least twelve yards. Once established they are topped at each weed cut. The weeds are then cut from bank to bank below the bar until the next bar is reached. Some keepers rake the gravel between the bars, this attracts the fish and on some waters the gravel when disturbed comes up very white, making a wonderful contrast. Don't leave too great a distance between bars, as trout and grayling feel safe if a refuge is near. It may be necessary on some waters with sharp bends and soft peat banks, to leave a further growth of weeds on the outer bends to prevent erosion. Inner bends should be cut.

Alternatively some keepers prefer the more difficult and time consuming chequer board style of weed bed management. As with line and bar cutting, it is only carried out on easily waded water.

Pole scythe

The pole scythe (134) is used to cut weeds in water that is too deep to wade. They are not commercially available and have to be made up. A three foot scythe blade is fixed to a long handle known as a pole, at an angle between 50-70 degrees. The fittings and stay are often from a redundant scythe. If both banks are to be cut this way, two will be needed, as the blade once set for one bank will be the wrong way around for the other. larch wood provides the best poles; larch thinnings of 16-18 feet are ideal, after drying out for a few months, they are prepared by removing all the bark and side shoots with a draw knife. The finished poles are surprisingly light, with a diameter of 2 inches tapering to less than ½ inch. Sometimes aluminium poles are used but these are a little heavy, and if they are not painted wear gloves to keep the tarnish from your hands. The pole scythe is held near the blade and thrown out into the weeds, the pole runs through the fingers and is then grasped at its opposite end. It is then pulled (never lifted) back towards the keeper with a jerking motion, cutting weeds as it is retrieved. A step is taken usually downstream and the operation is repeated. The pole scythe can be used successfully from a punt. Usually one person holds the punt on station while a colleague cuts the weeds. For safety reasons it can be dangerous to attempt boat cutting single-handed.

Chain scythe (links)

On deeper waters which can't be waded such as the middle and lower Avon, Test and Itchen, chain scything used to be the preferred method of clearing large areas of weeds. This style of cutting is particularly suited to the end of season weed cuts, when the river needs cutting hard to encourage winter scouring. The firmer upper-crust of mud and silt on slow flowing stretches can also be broken down with repeated cutting. Chain scythes consist of a number of blades that have been bolted together in such a way as they can pivot. The number of blades depends on the width of river. A keeper stands on each bank, ropes are attached to each end of the chain scythes. The blades are then pulled upstream with a short see-saw motion. Custom-made chain scythes were made years ago, and many are still in use today. The cutting edge of the blades were slightly angled and bevelled to reduce wear.

The Clearweed weed cutter has also been favoured by many fishing clubs. It consists of sprung steel blades that are twisted and serrated on both edges. Each section is thirty-six inches long and the standard length is made up to twenty-seven feet, although longer can be supplied. It is used the same way as a chain scythes and weighs around five pounds.

With the much reduced weed growth of recent years, the use of chain scythes has almost died out. Even 50 years ago there was much criticism of the excessive and indiscriminate use of chain scythes. Deeper waters are today frequently cut by a weed cutting boats.

Weed cutting boats

Before contemplating the use of a weed cutting boat (135), the user should be aware that there is strong evidence to suggest that the repeated use of weed boats on some sections may have an impact on channel depth and ultimately on the communities of plants present. The use of a boat should only be contemplated on water that is already deep and choked with ribbon and pipe weeds and where access for chain scythes is not feasible.

135 Weed cutting boats are used in deep water *Photo: Guy Robinson*

Weed cutting boats are expensive to buy, run and maintain but can cut large areas fast. They are launched and retrieved via a special trailer which has a break-back facility. Electric winches allow the boats to be hauled in and out of the water. A vehicle with off-road capability is also needed to tow the machines from place to place. A number of companies manufacture weed cutting boats. A traditional weed cutting boat as used by some private fisheries, will cut a swathe at least six feet wide. Some smaller boats can work in as little as twelve inches of water, which make them ideal for cutting carriers. Weed cutting boats are particularly suited for cutting wide and slow flowing reaches of water, especially hard cutting such as large areas of Common Club-rush. A number of contractors charge by the day for this service. Although expensive, in years of prolific weed growth the cost has to be weighed against overtime or the cost of casual labour. The July and August cutting periods are each usually only seven days. If a boat is hired for the more difficult large areas the river keeper is free to concentrate on beats that may need a more selective approach. Before booking a contractor, check the clearance under any bridges. He will also require a suitable area to launch and retrieve his boat, usually a firm bank with a gravel shallow. Advise downstream neighbours of the dates when a boat will be cutting your beats. Weed cutting boats are expected to finish cutting at least one day before the end of a cutting period to allow their cut weeds free passage down stream.

Weed jams

During weed cutting periods, areas where cut weeds may hang-up should be checked regularly. During stormy weather be particularly vigilant. If not tackled at once, phenomenal amounts of weed can collect and take hours to clear away. Correspondingly, footbridges can be pushed over by the sheer weight of accumulated weed. Vulnerable points such as narrow sluices, bridges, weed racks etc. often need checking last thing at night. Branches that hang in the water may hold up weed but they provide excellent cover for fish, so only a light trim should be considered. Footbridges should be cleared of hung weed regularly. Often a stout pole is left by each bridge for convenience.

Weed racks

Weed racks today are built to protect water courses that depart from a river with weed cutting activity. Cut weeds can choke a side stream if it is not suitably defended. Areas at particular risk would include inlets to fish farms, mills, cattle drinks and lakes etc.

In days gone by many estates in the middle and lower regions of some chalk streams, maintained large weed racks. They were sited on the main river as near to the top boundary of the fishery as possible, always on deep water. Stout larch poles were used in their construction, fixed in such a way that some poles could be removed allowing the river keepers to clean off hung up weeds. The origins of these large structures were in the days before co-ordinated weed cutting was organised. These weed racks were of a rugged construction, built to withstand the pressure of masses of cut weed compressed against them. Once weed cutting became unified some racks fell into disrepair, while others continued in use; typically they were cleared at the end of a weed cut and then weeds were allowed to build up until the next weed cut. This practice kept the rivers free of the nuisance of cut weed, particularly during wet weather, when a rise in levels can wash weeds down stream.

Herbicides

Chemical control of river weeds and bank-side herbage is a specialised activity. If contemplating this kind of treatment, it is vital that the Environment Agency is contacted in advance. If the length of river in question is designated as a SSSI, Natural England would also need to be notified.

Planting weeds

In the aftermath of the drought years 1989-92 many keepers planted water weeds in large areas that had been covered with blanket weed (*Cladophora*). *Ranunculus* is the most desirable weed and the one usually chosen for transplanting.

On some sections of river it may be a weir or an impoundment that is causing less than favourable conditions for weed growth with deep silt accumulating in the slow-flowing water upstream. Rather than planting weed in unsuitable conditions it may be best to remove such an obstruction. It is amazing to see how weed will grow if the conditions are right. A good example of how resilient *Ranunculus* can be is on winterbourne chalk streams which can remain dry for several years before springing back into life with luxuriant weed growth. If weeds do not come naturally there may well be problems with the substrate, water depth and velocity.

If you are contemplating weed planting, then the method used for gathering is to simply find a bed of young weed (well before flowering) and dig some of it up with a fork leaving the roots as undamaged as possible. It is then taken from the donor site and placed firmly in position without delay. In slow flowing areas weeds can often be planted by simply making a depression with your boot heel, or pinning the weeds to the river bed with flints. Probably the most consistently successful way, is the use of hessian sacks. These are part filled with some bottom material from the donor site and three slits are cut into one side, into which the roots of the weeds are pushed. The slits are then closed around the stems and stitched up. This is a particularly valuable method in quick water and eventually the bags will rot away, by which time the root system will have become established. It may be necessary to provide some additional anchorage. Success will largely depend on whether the river's habitat and flows are sufficient to support the transplanted weed. *Ranunculus* will need fast flowing water and good gravel habitat. River weeds must not be introduced from other river systems.

Trees

Trees are to be encouraged in areas that are farmed almost to the river bank or where chalk streams meander through open water meadows. Native species of trees not only enhance a river visually, but are vital to certain species of mayfly and they provide lies for fish. During prolonged windy conditions Mayfly (*Ephemera danica*) can be blown away from the water completely. Where trees are present, swarms of duns and spinners can be seen sheltering day and night in the safety of the branches. Trees play an important role in shading the channel and will help to keep water temperatures down during long hot spells. Often the very low scrubby cover that trails into the margins provides excellent cover for trout, particularly in low flow years when weed growth can be poor. Although trees are often viewed as a casting hazard by the angler, the river keeper has the very tricky job of balancing the needs of the angler with those of the fish.

Mowing

For many anglers much of the pleasure of having a rod on a chalk stream beat is the abundance of wild flowers and water meadow plants in the spring. Many plants, such as Water-docks, flags (Yellow Iris), Butterburs and various orchids are attractive to view and encourage insects which would not otherwise be found. In days gone by when the river banks had to be cut with scythes, mowing would be left until the grasses were waist high. With the advent of mowers, the operation was speeded up but the first cut was still often left until mid June due to the wetness of the banks. Today, using modern four-wheel drive mini-tractors, mowing can be undertaken earlier than would have been the case years ago. It is important that in areas that have these wonderful water meadow plants, the first mowing of the year is delayed to allow them to flower. Some owners compromise by just cutting a narrow path for anglers to tread around plants such as Ragged-Robin, Water Forget-me-not, and Lady's Smock.

Fishing instruction

Many river keepers are expert anglers in their own right and are more than happy to offer advice to a new angler on their beat. Expert in what is hatching that day, they will suggest the most appropriate fly and perhaps even where a prime fish can be stalked.

General Section

Poaching

Poaching salmon and trout from the chalk streams is a cause of concern to owners, keepers, and the Environment Agency who spend considerable sums of money on fishery protection. The responsibility for policing fisheries is generally confined to three authorities: the Environment Agency, whose Fisheries Department has its responsibilities and powers in relation to the legality of fishing controlled by the Salmon and Freshwater Fisheries Act 1975; the riparian owners' appointed agent, fishery manager, or keeper who can act as a private individual, without a warrant and is protected primarily by the provisions of the 1968 Theft Act; and the police, who can utilise any appropriate legislation and of course have a responsibility to protect persons and property.

The water bailiff

The Environment Agency appoints the water bailiff and issues the bailiff a warrant, the warrant being the instrument of his powers, which entitles him to:–

Demand the production of a valid fishing licence (as can any other licence holder).

Enter upon any land adjoining or near to a river within his authority's area, in order to inspect any obstruction, dam, weir, or fish trap.

Examine tackle, vehicles, boats and bait which he may suspect is being used, has been used, or is likely to be used for the illegal taking of fish.

Seize any equipment or vehicle if found to be involved in poaching activity under the Salmon and Freshwater Fisheries Act 1975.

The river keeper

Such person is of course appointed by the owner of a private fishery or his agent. His authority to turn trespassers off the land and remove any person fishing without consent, is restricted only to the area which constitutes his employer's property. He must use no more force then is reasonably necessary in removing any offenders. He has the right of citizen's arrest and seizure of tackle within the limits imposed by the Theft Act 1968, though it is advisable to involve the police to deal with such incidents. The keeper must make a reasonable demand for the offender to comply before resorting to the use of any physical contact and if he (the keeper) holds a current Environment Agency fishing licence, he may also demand the production of a licence from the offender.

Areas at particular risk are waters adjacent to rights of way, lay-byes, and road bridges. A new keeper should familiarise himself with these places as soon as possible. Where poachers persist in fishing from road bridges, occasionally keepers stretch a length of barbed wire across the river to stop trout being lifted out of the water.

The huge increase in rural crime over the past few years has encouraged many land owners to become actively involved in various farm watch schemes. River and gamekeepers often keep an eye on each others beats as they go about their duties. Larger farms and estates keep in touch with two-way radios. A popular target for theft is often anglers' cars, particularly when they are left beside isolated country lanes. Many estates and fisheries try and encourage anglers to park away from lanes often providing parking areas at a safer location.

Fishing clubs and private fisheries often hold trout in ponds for restocking during the fishing season. Poaching gangs will steal trout from such places causing considerable damage in the process. All grills should be fixed in such a way that their removal is impossible. Hatch boards that maintain water levels should also be secure to prevent the pond from being drained. Often stout pegs are driven into the bed of the pond in such a way as to snag any nets that may be used to steal trout. Fishery seine nets or landing nets should be kept well away from trout ponds to prevent them being used by poachers. There are specialist security firms that market various intruder alarms suitable for use in remote areas.

Professional keepers and Environment Agency bailiffs become adept at spotting poachers before the offenders realise they have been noticed. A keeper or owner must always assess the situation before approaching any suspected poachers. Never underestimate the lengths to which a gang will go to evade capture. If poachers are seen, the worst course of action is to rush in and deal with them without 'back-up'. Often it is better to retreat and if possible get the make, colour, and number of any car that you suspect they are using. Then call the police and tell them that poachers are fishing your water and give them the vehicle number and directions to your location. If it is practical, return to your water keeping a safe distance and gather what evidence you can. A description of any person taking fish will be useful in court. Frequently when the police arrive, poaching equipment and incriminating evidence is discarded or quickly hidden. In remote country areas the mobile phone can be a great asset when dealing with this type of offence.

The most persistent form of poaching today is gangs of youths usually in the eighteen to thirty year age group, who travel in cars from place to place fishing at random from road bridges. Frequently two people fish and one stays in the car, ready for a quick get-away. Typically hand-lines are used which consist of a length of line of about twenty pounds breaking strain attached to a short length of wood. The other end of the line has a hook and bread is used as bait. They are usually loath to stray far from their vehicles and can be particularly offensive when approached. When poachers of this type suspect they have been seen they usually drop the hand line in the river and make off.

Beat markers

Beat markers are used on fisheries to guide visiting anglers and guests around the beats. They are particularly useful when the keeper and the host are busy. After directions to the fishery, an angler should be able to find his way around his beat until the keeper is able to join him. Care must be taken to ensure that a guest is not left uncertain as to the extent of his beat. If for example, his beat is double bank fishing, a marker should be clearly visible on each bank. The top and bottom boundaries of the beat should be clearly defined.

Fishery rules

If a fishery is going to be let on a commercial basis, it is vital that a set of rules are drawn up. Visiting anglers should be sent or given a copy prior to starting to fish. Some fisheries have a copy displayed on the fishing hut notice board. Rules should ideally be as brief as possible. For example, if your fishing is half-water, you may wish to state 'no fishing past mid-stream'. Wading is allowed on some beats and not others; if it is dangerous to wade due to the nature of the river bed, visiting anglers should be made aware of this in the rules.

Typical points which rules should clarify:–

1. All anglers must hold a current Environment Agency licence.
2. Is nymph fishing permitted, from a particular date?
3. Is it dry fly upstream fishing only?
4. Hook size limit (some estates allow a larger size hook at mayfly time).
5. Barbless hooks only?
6. Is catch-and-release allowed?
7. Are dogs welcome?
8. Can two people share the rod?
9. Are coarse fish to be returned or killed?
10. How many trout can be taken?
11. Is there a minimum size?
12. How early are rods allowed on the water?
13. What time should fishing cease? (one hour after sunset is usual).
14. Is wading permitted?

Some fisheries also have a disclaimer in the rules which points out that there can be no liability if, for example. a fisherman's car is broken into, or any kind of accident befalls them while fishing.

Fishery records

Whether a fishery is a few hundred yards in length or a few miles, accurate returns of fish taken by anglers should be kept. Some estates have records of fish taken that go back to the 19th century. Because angling pressure on our rivers today is far greater than in years gone by, fishery returns are particularly relevant as an effective management tool. The best method is the return card system. One side of the card bears the fishery address, the other is divided into various headings, with boxes to be filled in giving weights and numbers of fish taken. A return card box placed by the fishery exit encourages anglers to complete the cards promptly, and save postage. Many fisheries have a return book in the fishing hut; this is useful to visiting anglers who can refer to the book for up-to-date information on catches of trout and successful flies used. Unfortunately there is always the risk of the book being taken or damaged if the hut is vandalised.

Equipment

Essential hand tools

Turk scythe
Pole scythe (if water is to deep to wade)
Mud-pan
Bow saw
Grapple
Hay knife
Trimmer (rip-hook blade attached to a handle around four feet long)
Grabs
Rakes
Slasher
Bill-hook
Crowbar
Sledge hammer
Post bumper
Stout round iron bar around six feet long
Two grain prong
Shovel
Spade
Fork
Ropes
Turfer winch

Winches

When trees collapse into the river it may be necessary to recover them in areas that are inaccessible to heavy vehicles. Under such circumstances, a portable hand winch can be invaluable. They are also used for extracting deeply driven piles from the river bed. Obviously they rely on the convenient location of another tree or anchor point.

Portable electric winches

Portable electric winches are becoming increasingly popular for such tasks as hauling falling trees to the riverbank, straining fencing and the launching and recovering of boats. One manufacturer markets a model which fits straight onto the ball hitch of any vehicle. They are powered from a standard car battery and have a line pull up to three tons. A useful safety feature is the remote control hand-held unit, which enables the user to stand aside from the direct line of pull.

Grass cutters

These machines fall into three categories: ride on, self-propelled and hand-propelled. Many fisheries and estates use mini tractors for mowing because of their great versatility throughout the year. If contemplating the purchase of such a machine, remember that the nature of the river bank that you intend to mow will influence your choice of tractor and cutting unit. If your river banks are soft, oversized tyres and four wheel drive would be an important consideration. Mid-mounted rotary cutting units are popular and efficient on reasonably firm and level banks. They are unsuitable on soft undulating banks because of their poor ground clearance. In areas such as these, a rear-mounted mower driven by the power take-off and held in place by the hydraulic linkage would be more suitable. Once you have an idea of your requirements, ask for a demonstration on your water before purchasing. Check gateways and bridges that you intend the machine to use for width.

Brush cutters

Brush cutters today have largely replaced the mowing scythe for much of the mundane work around a fishery. Modern two-stroke brush cutters are light and versatile, and come with three main attachments. The most useful is the ridged blade, which is used to clear paths and cut back reeds and sedge. The brush cutter's weight is supported by a harness, which usually has a quick-release button for safety. The action used by the operator is similar to when using a scythe. The blade can be removed and re-sharpened in a vice with a suitable file. Always check carefully before refitting for any cracks in the metal, and discard if any are found.

The mowing head attachment has usually two lengths of nylon which produce a particularly tidy finish. It is useful for mowing around trees and seats without fear of damaging them.

The brush knife is suitable for cutting back scrub, brambles and matted grasses. The blade can also be sharpened.

Brush cutter safety

A brush cutter can be a particularly hazardous tool if safety precautions are not followed. Careless and improper use could cause a serious injury. Anyone using a brush cutter should be in good physical condition. Always ensure bystanders, children and animals are not allowed in the area. It is vital that the appropriate safety clothing is worn at all times when using a brush cutter. This would include a face screen or goggles, ear protectors, steel toe-capped boots and non-slip gloves.

Chain saws

Today, most farms, large gardens and fisheries would number a chain saw amongst their machinery. Usually the semi-professional, heavy-duty models are a sound investment. If contemplating buying a second-hand model, pay particular attention to the bar, chain and sprocket. If one of these items are worn, there is a good chance they all will be. Many older saws are not fitted with a chain brake. It is now illegal to issue such a saw to employees. Although the chain saw enables the river keeper to get heavy work done quickly, it also presents a grave source of danger. Too many operators are maimed in a single moment of thoughtlessness. There is available a full range of protective clothing designed specifically for use with chain saws, and every operator must take advantage of this important innovation.

The National Proficiency Tests Council run excellent courses on such items as chain saw maintenance and safety, felling, de-limbing and getting down hung up trees. Certificates of competence can be awarded and gradually a trainee can become competent in a range of chain saw skills. Never use or expect others to use a chain saw without proper training. Employers have a legal obligation to ensure that their employees who use a chain saw have the appropriate protective equipment and have completed the relevant training.

136 A chainsaw being used to cut willow faggots *Photo Guy Robinson*

The workshop

Anyone running a fishery today will save time and money by having their own workshop. Ideally, the workshop should provide a dry, well-lit and ventilated area where tools can be repaired or adapted for future use. It is often helpful if partly completed work can be left safely with the knowledge that it will not be interfered with. Any workshop should be sited near to a house for security reasons, and preferably have a secure area within the workshop where power tools can be locked away safely. The level of equipment can be built up slowly but some items, such as a work bench and a large vice, are essential. Basic tool kits for a carpenter, engineer and mechanic will also be invaluable.

References

Bingham, C. *The River Test, Portrait of an English Chalkstream,* H.F. & G. Witherby, (1990).

Giles, Dr N. and Summers, Dr D. W. *Helping Fish in Lowland Streams*, Game Conservancy Ltd. (1996). ISBN 0 9500 130 9 9

Giles, Dr N., *The Nature of Trout* (2006), available via: www.percapress.com

Hendry, Dr K. and Cragg-Hine, Dr D. *Restoration of Riverine Salmon Habitats – A Guidance Manual.* Environment Agency R&D Tech Rep W44 (1997).

Lewis, V., *The Wild Trout Survival Guide*, Wild Trout Trust, Environment Agency, River Restoration Centre (2006), available from www.wildtrout.org

Lunn, M. and Graham Ranger, C. , *A Particular Lunn: 100 Glorious Years on the River Test,* Allen & Unwin (2006).

Mainstone, C. P. *Chalk Rivers Nature Conservation and Management*, English Nature and Environment Agency (1999).

O'Grady, Dr M., *Channels & Challenges – The Enhancement of Salmonid Rivers*, Irish Central Fisheries Board (2006) www.cfb.ie

Seymour, R., *Fishery Management and Keepering,* Charles Knight & Co. (1970).

Summers, Dr D.W., Giles, Dr N., and Willis, D. J., *Restoration of Riverine Trout Habitats – A Guidance Manual.* Environment Agency R&D Technical Report W18 (1996).

Talks, L., Weatherley, Dr N., Frake, A., Mainstone, C., *The State of England's Chalk Rivers*, Environment Agency (2004).

Templeton, R. G., *Freshwater Fisheries Management*, 2nd Edition, Fishing News Books (1995).

Ward, D., Holmes, N., and Jose, P., *The New Rivers and Wildlife Handbook,* Sandy, Bedfordshire (1994).

RETROSPECT

'The beauty, however, of chalk-strewn valleys still remains wonderful.' The river still waters meadows that are unspoilt and unchanged, and its clear purity is guarded and protected.

Still glides the stream and shall forever glide,
The form remains, the function never dies.'

From the concluding chapter of Fly Fishing *by Sir Edward Grey, later Lord Grey of Fallodon., K.G. Published in 1930 by J.M. Dent & Co.*

137 Upper Test at Polhampton

138 Itchen water meadows at St Cross

139 Upper Test at Tufton

140 Middle Test at Bossington

Photos: R H Dadswell